Structural Analysis of Thermoplastic Components

Structural Analysis of Thermoplastic Components

Gerry Trantina, Ph.D.
Ron Nimmer, Ph.D.

Edited by
Peggy Malnati

McGraw-Hill, Inc.

New York   San Francisco   Washington, D.C.   Auckland   Bogotá
Caracas   Lisbon   London   Madrid   Mexico City   Milan
Montreal   New Delhi   San Juan   Singapore
Sydney   Tokyo   Toronto

Library of Congress Cataloging-in-Publication Data

Trantina, Gerald G.
   Structural analysis of thermoplastic components / Gerry Trantina,
Ron Nimmer ; edited by Peggy Malnati.
     p.   cm.
   Includes bibliographical references and index.
   ISBN 0-07-065202-3
   1. Heat resistant plastics. 2. Structural analysis (Engineering)
I. Nimmer, Ron. II. Malnati, Peggy. III. Title.
TA455.P5T75 1994
620.1'9231—dc20   93-35994
                                              CIP

1 2 3 4 5 6 7 8 9 0 DOC/DOC 9 9 8 7 6 5 4 3

ISBN 0-07-065202-3

*The sponsoring editor for this book was Harold B. Crawford, the
editing supervisor was Valerie L. Miller, and the production supervisor
was Pamela A. Pelton. It was set in Century Schoolbook by Carol
Woolverton, Lexington, Mass.*

Printed and bound by R. R. Donnelley & Sons Company.

Cover photos courtesy of GE Plastics.

This book is printed on acid-free paper.

Contents

Acknowledgments

Since the early 1980s, General Electric Corporate Research and Development (GE-CRD) has fostered an active, growing program in developing technology for the structural design analysis of thermoplastic parts. That long-term commitment by an organization as well as the dedication of the individuals associated with the technology development is what makes this book possible. Our colleagues on that team—Vijay Stokes, Horst deLorenzi, Herm Nied, Joe Woods, Andy Poslinski, Wit Bushko, Lou Inzinna, Harry Moran, Ken Conway, and Linda Briel—have all made significant contributions to the technical understanding that forms the basis of this book. We would like to especially acknowledge Vijay Stokes for his dedication and accomplishments in the area of mechanics of plastics, as well as his demonstrated leadership and creativity in maintaining the focus of this long-standing program. Although we are engineers by training, technology development in an area such as this takes place most effectively when engineers and material scientists have the opportunity to collaborate. Fortunately for us, our team benefited from the existence of an even more long-standing team of scientists working at GE-CRD in the area of polymer chemistry and physics. Collaboration and discussion with them, especially Albert Yee (now at the University of Michigan), Mike Takemori, Dean Matsumoto, Roger Kambour, Don LeGrand, and Stan Hobbs, have greatly enhanced the mechanical engineering technology development represented in this book.

The success of any technology development is measured, in part, by the degree to which it is applied in a commercial environment. Significant contributions have been made by a large number of mechanical engineers with whom we have had the privilege to work. Some, including Todd Paro, Laurie Miller, Dave Ysseldyke, Orville Bailey, Ken Sherman, Doug Wright, Tim Craven, Janet Rawson, Charley Taylor, Dave Kazmer, Pete Oehler, and Todd Hoff, not only helped to develop structural design analysis techniques for thermoplastic parts at GE-CRD, but then also implemented and applied the technology as application engineers at GE Plastics. Other mechanical engineers at GE Plastics, including Mark Minnichelli, Pete Zuber, Dan Furlano, Ray Kolberg, Greg Ambur, Ron Domingue, Larry Doheny, Phil Richards, and Pete Maruk, not only effectively applied and improved new technology, but also helped identify the next generation of technology issues. Many thanks to GE Plastics for furnishing cover and other photos to illustrate this book.

In addition to the conscientious dedication of scientific and engineering personnel, sustained technology development requires the vision and patience of business leaders and technologists. Ten to 15 years ago, few people saw any need to invest in technology that would be necessary for plastics to fill the role of a load-bearing engineering material. Fortunately, enough of those leaders surrounding us, including Roland Schmitt, Joe Wirth, Don Mowbray, Ed McInerney, Bill Schlich, Mike Brown, Piet van Abeelan, Jean Hueschen, Tom Olliver, Jack Avery, and Dick Lassor, did recognize the importance of engineering analysis and design technology for plastics. Without their avid support, little of this work could have been accomplished.

Finally, very special thanks must be conveyed to Julie Kinloch for her patience, dedication, and skill in preparing the manuscript. Engineering technology would progress much more rapidly if all engineers were as efficient as Julie. We dedicate this book to our families—Sharon, Cynthy, Aaron, Sarah, Jeff, Mike, and Ken—for giving up family time so that this effort could be completed. And lastly, "Here's to you Mrs. Robinson."

Structural Analysis
of Thermoplastic
Components

1

Introduction

Purpose

Engineering design requires both mechanical properties to define material behavior adequately and accurate analysis techniques to predict generic part performance based on those data. The primary purposes of this book are to assess the current effectiveness of that technology with respect to thermoplastic materials and to suggest improvements to enhance engineering analysis of thermoplastic parts in the future. Increasingly, plastic materials are being considered for use in load-bearing components, and the ability to apply mechanical analysis effectively to design for performance continues to grow in importance. In order to foster this technology growth, issues specifically relevant to the mechanical behavior and analysis of plastic parts must be identified, approaches for handling these issues defined, and experience in their effectiveness documented.

Despite their large volume of use, polymers are still a relatively new class of engineering materials. In fact, the use of the term *polymer* dates back only to 1832 when it was applied by the Swedish chemist Berzelius.[1] Whenever a new class of materials enters engineering use, there is a necessary period of technology development and adjustment before a well-structured and logical process of design with such materials reaches maturity. Before such a logical process can evolve, several much more fundamental developments must occur. For example, elemental structure of the material must be well understood in order to facilitate material invention and fundamental evolution. Furthermore, the processes used for forming the material into useful shapes must become routine and well controlled, and an understanding of the relationship between the material's morphology and its mechanical properties must also develop. As these technologies mature, it becomes possible to recognize whether the material's properties and manufacturing processes offer specific advantages for achieving the functional requirements of a part. Out of these technologies arises a framework for conceiving potential approaches for application of the material and manufacturing processes to achieve a functional goal.

If, after sufficient development of these fundamental technologies, materials evolve with mechanical properties sufficient to consider their use in load-bearing applications, then engineering design will be required to determine the necessary size and shape of the part to achieve the required level of performance. That process is greatly facilitated if the material properties necessary to define engineering performance of a component are well defined and there are accurate analytical procedures available for application of these properties to define geometry. Once these quantitative methodologies are in place, the ability to understand mechanical behavior and to identify the most effective use of the material is significantly enhanced. Without their presence, the only alternative is a time-consuming and costly approach of trial and error.

At present, the fundamental groundwork in chemistry, materials science, and process engineering for polymeric materials is well in place. There have been numerous books written treating all of these subjects.[2–14] In addition, there has also been significant discussion in the literature devoted to the geometric definition of plastic parts to ensure successful processing and fabrication. For many applications, this information may be sufficient. However, when a component must fulfill a function of bearing load as part of its application, the necessity of defining geometry to guarantee performance becomes an additional requirement for success. As will be seen from the brief historical perspective that follows, the general state of development of plastic materials and their engineering application has now evolved to a point where the analysis of mechanical performance merits and will experience more attention.

Historical Perspective

The first plastic to be produced commercially was Celluloid™ polymer. The history of this material can be traced back to written records[15] attributed to J. Pelouze in 1835 and the German chemist Schonbein in 1845 that document observations of an interesting new material derived from mixing wood fibers or paper with concentrated nitric acid. In spite of the lack of understanding of the basic chemistry of this material, a 19th-century English materials technologist named Alexander Parkes was able to initiate commercial production of a material he called *Parkesine,* which was related to the earlier recordings of Pelouze and Schonbein. Parkes had no formal education in chemistry but had experienced considerable success in processes related to both metallurgy and rubber. He exhibited his material at the Great International Exhibition of 1862 in South Kensington, England and was awarded a bronze medal for excellence in quality. Visitors to the exhibition were impressed with the low cost and formability of this new material as well as its hardness once completely processed. In 1866, Parkes launched the Parkesine Company, an enterprise to use the Parkesine polymer for production of commercial products such as combs, umbrella handles, tableware handles, and jewelry. However, for a variety of reasons including extreme efforts to minimize production cost, the quality of these products was less than acceptable and the company was liquidated in 1868.[15]

During this same period of time, another imaginative inventor in the United States named John Wesley Hyatt had started pursuing the production of a very similar material. Like Parkes, Hyatt had no formal education in chemistry. As a result of a growing shortage of ivory, the billiard-ball manufacturing enterprise of Phelan and Collander in Albany, New York had offered a $10,000 prize for the invention of a replacement material for the ivory in their product. It was in pursuit of this prize that Hyatt initiated his investigation. Hyatt's first patents in this area appeared in 1865,[15] but it was not until 1870 that the patent covering his discovery of the contribution of camphor in processing celluloid appeared. This discovery eventually allowed Hyatt to solve most of the problems that had plagued Parkes in England. In 1872, Hyatt and his brother formed the Celluloid Manufacturing Company and embarked on a very successful commercial introduction of this material and products made from it. The business grew and eventually moved to Newark, New Jersey. Its name evolved to the American Celluloid and Chemical Corporation and it was eventually absorbed by the Celanese Corporation, which has more recently been acquired by Hoeschst to become Hoeschst Celanese.[15] The commercial plastics industry had achieved its first success.

In spite of the success of Celluloid® polymer as a material, other plastic materials did not rapidly appear. The continued lack of understanding regarding the basic structure and chemistry of this new class of materials proved to be a major impediment to the introduction of additional materials of a similar nature. In fact, it was not until 1877, five years after the establishment of Hyatt's business, that the existence of long chains of molecules was hypothesized by Kekule[16] as relevant to the very special properties associated with certain natural organic substances like cellulose.

It was approximately 41 years after the discovery of Celluloid polymer that the second plastic material was developed by Dr. Leo Bakeland in the United States and named Bakelite® polymer.[17] Although Bakeland had developed the material as a synthetic electrical insulator, John Hyatt, who had also founded the Hyatt-Burroughs Billiard Ball Company, applied it to billiard balls once again. While Celluloid polymer was the first plastic, it was really only semisynthetic since it was made of natural materials. Bakelite polymer was the first truly synthetic plastic.

The introduction of new plastic materials still did not accelerate with the discovery of Bakelite polymer. By 1930, almost 60 years after the commercial introduction of the first plastic, the annual production of these materials in the United Sates was only 23,000 metric tons, almost all of which was Celluloid and Bakelite[16] polymers. A fundamental understanding of the basic chemistry relevant to plastics was still the most significant technological issue impeding progress toward new polymeric materials with a wider range of potential applications. That situation was soon to change.

The scientist who is generally honored for having led the way in founding polymer chemistry is Hermann Staudinger.[15] Staudinger wrestled with basic issues relevant to the size of the molecules of which natural and synthetic polymers were made and the character of the forces that held them together and gave them their peculiar properties. In contrast to the majority of investigators dealing with such issues in the 1920s, Staudinger did not subscribe to the concept that these materials consisted of small molecules, clustered in aggregates and held together by some electrical or yet-to-be-defined force of molecular attraction. Instead, Staudinger was a proponent of the idea that the atoms comprising polymers were bound together by the same covalent bonds understood to exist in normal organic compounds. His experiments led him to believe that the only significantly distinctive feature of polymeric materials was the immense size of their molecules. In 1924, Staudinger proposed that polystyrene and natural rubber had linear structures with extremely large molecular weights—about 20,000 for polystyrene—made up of identical, repetitively linked, basic chemical units.[15] He went on to explain that unlike other compounds such as water, which has a fixed and identical molecular weight associated with all its molecules, the molecular weight of a polymer was merely an average. In 1953 Staudinger received the Nobel prize for his pioneering work in polymer chemistry, but his work in understanding the basic structure of polymers helped fuel the expansion of polymer technology well before his award.

As the molecular structure of polymers became better understood, there was also progress in defining mechanisms by which these polymers could be produced. The work of Dr. Wallace H. Carothers was particularly important in this area. Carothers's contributions to the science of polymer synthesis includes emphasis on the concept of "functionality" in polymeric reactions as well as the distinction between the two important classes of polymerization: namely addition and condensation. Beyond his fundamental contributions to polymer science, Carothers is also remembered for his management of the Du Pont Company research laboratory, where he invented the synthetic rubber, neoprene, as well as the thermoplastic polymer, nylon.

The next impetus in the development of polymer technology was rooted in international history as much as in science: during World War II it was necessary to find new synthetic materials to replace inaccessible natural resources such as rubber. Coupled with the newly acquired knowledge of polymer chemistry, these immense new demands created the real surge in plastic materials invention.[15,17] Beginning in the years just prior to World War II, a number of today's familiar plastics began to appear. Many of the first applications were related to the outstanding electrical insulation properties of these new materials. Polyvinyl chloride, for example, was widely used as cable insulation. The dielectric properties of polyethylene made it another attractive material for use in electric equipment, and it became especially critical in the radar used by the English during the Battle of Britain. Polystyrene not only saw use in electrical equipment, but it was also associated, in one form or another, with the artificial rubbers developed for use as tires for most of the military vehicles. Clear and light in weight, polymethyl methacrylate was used extensively for cockpit canopies. Nylon, on the other hand, saw application in a very different form. Drawn into fibers, it was heavily used for parachute material, tow ropes, and tire chords. The technology and practical application of polymeric materials had obviously achieved a new and much more rapid growth rate. By 1949, the annual production of plastic in the United States had reached 517,000 metric tons (570,000 tons).[16]

After World War II, the rate of discovery and introduction of new polymers continued to accelerate. This growth helped stimulate the development of other technologies that were essential to the effective use of plastics. Processing technology in particular had to improve to meet the efficiency demands of a peacetime economy. The original Celluloid products were manufactured with a variety of techniques including injection molding and extrusion. However, the limited use of plastic did not merit any extensive work to refine these processing methods. That situation changed with the rapid introduction of new materials during the 1930s and 1940s. As material suppliers began looking for new peacetime applications for their plastics, there was new impetus for improving processing technology as well. Some of the new polymers were difficult to process, with narrow windows of temperature available for molding parts. Commercial efficiency became an increasingly important goal as the new plastics competed with older materials as well as each other for market share. Innovations in the existing processes such as preplasticizing systems began to appear rapidly. Although injection molding and extrusion remained the highest volume plastic processing methods, new techniques such as blow molding also appeared.

Another technology that began to develop and mature in response to the rapid expansion of the plastic industry was polymer materials science. Materials science might be defined as focusing upon the relationship between material properties and material form and structure. One example of the significance of this technology is the effect of molecular weight on plastic material properties. Reduced molecular weight is associated with lower viscosity at processing temperatures and can often make a plastic easier to mold. However, reducing a polymer's molecular weight also has negative effects on the impact resistance of the polymer, which in turn could make the performance of the molded component unacceptable. Understanding and controlling such effects is crucial to the success of a plastic component. Materials science has also helped industry come to understand that modifying or manipulating any one property usually also affects other properties. For instance, the addition of small rubber particles to the polymer will improve impact resistance, especially at lower temperatures when materials are more stiff. However, the presence of those particles also affects the flame-retardant properties of the material. The more rubber that is added to increase impact performance, the lower the fire-resisting properties of the polymer. And generally, the greater the amount of flame-retardant additives in the polymer matrix, the poorer the resulting material's impact resistance will be. Progress in the materials science technology area has also been very influential in expanding the application opportunities for plastic materials.

At this point, it is worthwhile to mention one additional element of the history of plastics. Although seemingly insignificant in comparison to some of the previously mentioned events, it is highlighted because of its significance with respect to the development of mechanical technology for plastics. In *The Story of the Plastics Industry*,[17] published by the Society for the Plastics Industry (SPI) in 1977, the appearance of a subgroup of plastics identified as "engineering thermoplastics" are mentioned during a brief discussion of relevant events occurring during the 1950s. Acetal (introduced in 1956) and polycarbonate (1957) are listed as examples, along with nylon which was developed much earlier. In describing the nature of this plastic subgroup, SPI singles out superior impact strength and thermal and dimensional stability, which allowed these materials to compete more closely with metals in load-bearing environments. Many other plastics that fall into this somewhat ambiguously defined subgroup have appeared since. This material evolution placed new pressure upon mechanical technology to meet the challenge of performance in structural applications. Furthermore, since these materials generally cost more than other plastics, there was also new pressure for mechanical and processing technology to be more efficient in the use of these materials.

With a solid basis established in polymer chemistry and rapidly evolving technologies in polymer processing and materials science, the annual production of plastic in the United States reached 2,476,601 metric tons (2,730,000 tons) in 1959 and swelled to 17,263,420 metric tons (19,000,000 tons) by 1978. In 1979, production of all plastics in the United States surpassed the production of steel on a volume basis.[16] At this stage in the development of plastic as a material, mechanical technology began to play a more significant role, continuing the success and expansion of plastic technology. Effective mechanical engineering and design were not always an important ingredient for success in many of the first applications of plastics that have been mentioned. It is not necessary to design a billiard ball. No geometry or structure must be defined for insulation used in an electrical cable. However, from their rather modest initial applications, plastic materials have evolved toward much more demanding applications that require sophisticated engineering analysis and design. For example, plastics are now being used to manufacture automotive bumpers, such as that shown in Fig. 1.1, which must be able to survive 8-km/h (5-mi/h) impacts without damage and without deflecting so much that other parts of the car are damaged. Satellite dish antennas, such as the one shown in Fig. 1.2, are being manufactured in plastic and must be designed such that they can withstand high winds without damage or distortion of their parabolic shape. In applications such as these, geometries

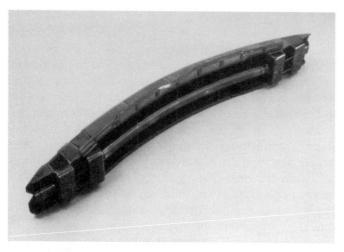

Figure 1.1 Plastic automotive bumper.

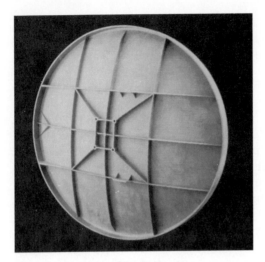

Figure 1.2 Plastic satellite dish antenna.

must be defined to fulfill mechanical performance requirements. In order for geometric definition to be most efficient, it must be possible to predict mechanical performance as a function of geometry based on the fundamental mechanical properties of the materials being considered. If this is not possible, then the mechanical design process can only fall back on the time-consuming and inefficient trial-and-error or build-and-test process—something that today's highly competitive and global economy can rarely cost justify.

Text Outline

In focusing on the quantitative mechanical analysis of plastic part performance, this text identifies issues specifically relevant to plastic materials, suggests practical approaches to handle these issues, and documents experience in real plastic part applications. It does not attempt to outline a general procedure for design of plastic parts. Instead, it concentrates on the specific task of using pertinent material properties to predict performance in general part geometries. And while material properties are discussed and applied, this is not a materials science text. There are already a number of books that treat that subject. Furthermore, although advanced analysis techniques (the finite-element technique in particular) are applied in many of the examples, the theoretical development of these techniques will not be emphasized. This subject is amply covered in other texts as well. Discussion instead will be limited to a level necessary to relate the techniques' use to the prediction of the mechanical behavior of plastic parts. Quantitative mechanical analysis approaches will be applied to illustrate significant issues in the design of plastic components and demonstrate their current level of effectiveness.

Part of the process of making efficient use of a material is accomplished by recognizing the peculiarities of that material—both its advantages and its disadvantages. Once recognized, the effects of these characteristics on component behavior can be investigated and approaches to maximizing advantages and minimizing disadvantages can be developed. There are a number of fundamental characteristics associated with plastics that are important to the process of mechanical design. For example, most plastics have very low Young's moduli—about two orders of magnitude less than those of many engineering metals. The ductile plastics also have low yield strengths—on the order of one decade less than metals. In contrast, the yield strains of plastics are much higher than those of metals—approximately one order of magnitude higher. Ductile plastics also may exhibit much larger strains to failure, sometimes approaching several hundred percent strain to failure. Although metals also exhibit temperature and rate sensitivity at high temperatures, these effects must be considered for many plastics at much more moderate temperatures. During the course of this text many of these characteristics will be examined and their effects on mechanical performance and design quantified and illustrated.

With historical perspective on the technology associated with polymers provided in this introductory chapter, Chap. 2 will focus on a number of the most basic elements and terms associated with chemistry, material categorization, and production processes. Although these subjects will not be treated in detail, some understanding will be useful as they relate to issues of mechanical design and analysis.

Chapter 3 presents a brief discussion of the finite-element analysis technique that is prevalently used today for structural analysis. Since there are a large number of texts that already treat the details of this technique, this chapter is intended to provide a general introduction to the method. In addition to an overview of the theory and application of this technique to linear problems, several nonlinear issues are discussed in this chapter because of their particular significance for plastics. These topics will be highlighted and discussed specifically with relevant examples.

Chapters 4 through 8 deal with five specific subjects relevant to mechanical analysis and design: stiffness, strength and failure, impact, time-dependent response, and fatigue. The subject matter in each of these chapters can be differentiated with respect to the load applied to the structure. In the stiffness and strength chapters, the loads that will be considered are primarily static in nature. These loads will generally be assumed to be slowly applied and then removed over a time roughly equivalent to the time during which they were applied. In contrast, the loads considered in Chap. 6, dealing with impact, are applied and re-moved much more rapidly, as illustrated in Fig. 1.3a. The question of what constitutes fast or slow loading rates will be addressed. Chapter 7, on time-dependent response, is associated with loads that are ap-plied and remain in place at a relatively constant level over time, as shown in Fig. 1.3b. Shown in Fig. 1.3c is the oscillating or cyclic load, which is relevant to the subject matter of Chap. 8, fatigue.

Stiffness, the load required to produce a unit of displacement, is af-fected by both material and geometry. Chapter 4 discusses issues rele-vant to both of these subjects. With respect to material influence, the material properties necessary for adequate prediction of structural stiffness are identified and discussed. For some plastic materials this is

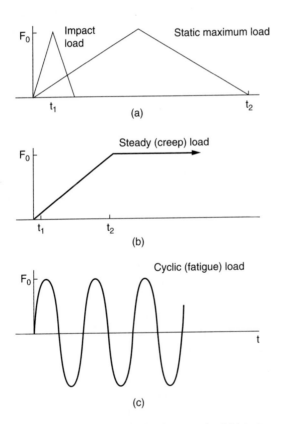

Figure 1.3 (a) Generic static and impact load histories. (b) Constant load. (c) Cyclic load.

a straightforward process. However, in other cases, such as structural foam and various glass-filled materials, the issue of adequate material property data for stiffness definition is more involved. Geometry will also play a role in the stiffness of a plastic part. Since the material stiffness of polymers is generally much less than that of other engineering materials, geometry must often be used to produce adequate structural stiffness. Examples of how this is commonly accomplished are presented and significant issues relative to the performance of such structures, as well as accuracy of prediction, are also discussed for simple components.

Beyond stiffness, the next most common constraint upon mechanical design is failure. The definition of failure is dependent upon component performance requirements. It may be based upon displacement or damage considerations. In Chap. 5, a number of different damage and failure mechanisms common to thermoplastics are discussed. Approaches to modeling and predicting this behavior are outlined and comparisons with experiments are offered.

In Chaps. 6 through 8, three generic areas of mechanical response technology—impact, time-dependent response, and cyclic loading—will be developed separately. Chapter 6 presents special considerations regarding the analysis of thermoplastic components designed to withstand impact events. This includes inertial effects in impact problems, rate-dependent material response, failure, and common material tests. Time-dependent part performance such as creep and stress relaxation is presented in Chap. 7. A simple model and theories, design analysis for time-dependent deflection, variable loading, recovery, and rupture are emphasized. Cycle-dependent part performance (fatigue) is included in Chap. 8 with emphasis on understanding important loading variables, component lifetimes, frequency effects and hysteretic heating, and cyclic crack growth.

It is impossible to discuss the behavior of all the currently available plastics relative to these engineering performance issues. Consequently, we will focus upon generic concepts related to mechanical technology of plastics in a broad sense. This approach will introduce some limitations in scope that should be defined now for clarity. The practical examples and comparisons between prediction and measurement that are presented are a very substantial element of this text and its goals. They provide real illustrations of general issues and illustrate attainable levels of accuracy with engineering tools. In providing realistic examples of the mechanical response and analysis of plastics, we will use materials with which we have the most experience. In most cases, these polymers will be thermoplastic in nature. It is not our intent to make any suggestion or recommendation with respect to material use as part of this process. Our specific choices are simply made as a matter of necessity to allow us to discuss issues with maximum background in knowledge and experience.

References

1. Herbert Morawetz, *Polymers: The Origins and Growth of a Science,* Wiley, New York, 1985, p. 3.
2. K. J. Saunders, *Organic Polymer Chemistry: An Introduction to the Organic Chemistry of Adhesives, Fibres, Paints, Plastics and Rubbers,* Chapman and Hall, London, 1973.
3. Walter Driver, *Plastics Chemistry and Technology,* Van Nostrand Reinhold, New York, 1979.
4. Reginald L. Wakeman, *The Chemistry of Commercial Plastics,* Reinhold, New York, 1947.
5. *Handbook of Polymer Science,* Vol. 1, *Synthesis and Properties,* Nicholas P. Cheremisnoff, Ed., Decker, New York, 1989.
6. *Handbook of Polymer Science,* Vol. 2, *Performance Properties of Plastics and Elastomers,* Nicholas P. Cheremisnoff, Ed., Decker, New York, 1989.
7. *Handbook of Polymer Science,* Vol. 3, *Applications and Processing Operations,* Nicholas P. Cheremisnoff, Ed., Decker, New York, 1989.
8. Arthur N. Birley, *Plastic Materials: Properties and Applications,* Chapman and Hall, New York, 1982.
9. J. A. Brydson, *Plastic Materials,* Butterworths, London, 1989.
10. I. M. Ward, *Mechanical Properties of Solid Polymers,* 2nd ed., Wiley, Chichester, 1971.
11. C. B. Bucknall, *Toughened Plastics,* Applied Science, London, 1977.
12. Stanley Middleman, *Fundamentals of Polymer Processing,* McGraw-Hill, New York, 1977.
13. Nam P. Suh and Nak Ho Sung, *Science and Technology of Polymer Processing,* MIT Press, Cambridge, MA, 1977.
14. *Plastics and Polymer Processing Automation,* Plastics and Rubber Institute, Noyes Data, Park Ridge, NJ, 1987.
15. M. Kaufman, *The First Century of Plastics,* The Plastics Institute, London, 1963.
16. Henri Ulrich, *Introduction to Industrial Polymers,* Hanser, Munich, Germany, 1982.
17. Joel Frados, *The Story of the Plastics Industry,* Society of the Plastics Industry, New York, 1977.

2

Materials, Processing, and Design

The development of a structural plastic part requires an understanding of the interrelationships of materials, processing, and design. In this chapter, information will be presented on plastic materials, processing knowledge, and the engineering design process, with the goal of providing background information for a design engineer who is challenged to structurally design a plastic part. This background information will also serve as a useful reference for terminology regarding specific materials and processes discussed later in this book. Plastics are complex materials with various mechanical properties that influence part performance. Reinforcements that are commonly added to polymers can affect these properties as can the process of blending and alloying different polymers to form new materials. Standard datasheets do not provide properties that are directly useful for predicting the structural performance of plastic components. In addition to the effect that a material's properties can have on design, the constraints and advantages of different processes can also play a design role. Design engineering involves definition of component geometry that, with the optimum selection of material and process, meets the end-use requirements at the lowest cost (Fig. 2.1). The presentation of material, process, and design engineering information in this chapter provides some useful background for the discussions of structural design analysis techniques covered throughout the rest of this book. More extensive general information about polymers, processing, and basic design are included in Refs. 1 to 6.

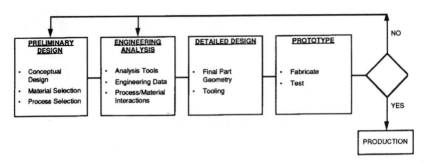

Figure 2.1 The goal is to meet the end-user requirements the first time with the lowest cost through the design engineering process.

Plastic Materials

The vast majority of plastics are synthetic compounds having a carbon–carbon backbone modified by other organic side groups. Plastics are differentiated chemically from other materials by their extremely long, chainlike molecules and high molecular weights. The process of joining together tens of thousands of smaller molecular units—known as monomers—into these long molecules is called polymerization. A plastic material is a polymer plus reinforcements, fillers, and additives, such as mineral and glass fillers, flame retardants, and stabilizers. The resultant plastics are then shaped and solidified, thus producing a plastic part.

In general, plastic materials can be divided into two major classes: thermoplastics and thermosets. Celluloid resin, the first commercial plastic, was a thermoplastic polymer. Bakelite resin, the second type of plastic to appear, was a thermoset. The fundamental difference between these two types of plastic materials can be described with respect to the bonds that exist between the long molecular chains and their consequent response to temperature increase. When raised to a sufficiently high temperature, a solid thermoplastic will become viscous and pliable in nature. In this state, it can be formed into useful geometric shapes and will retain these shapes when cooled. If reheated to a high enough temperature, a thermoplastic will again become viscous and can be reshaped and then cooled and resolidified. This characteristic behavior is related to the fact that the long molecular chains of thermoplastics are not significantly cross-linked through covalent bonds, similar to those bonds that hold each chain together. On the other hand, the first application of sufficient heat or a catalyst to a thermoset initiates covalent bonding between these polymer chains, leading to a more rigid, cross-linked coupling between the constituent molecules. Before this cross-linking occurs, a thermoset can be very

fluid in nature, allowing it to fill molds defining the desired component's shape easily. However, unlike thermoplastics, if a thermoset is
subsequently reheated, it will not return to its fluidlike shape. Sufficient application of temperature will eventually degrade the thermoset.

This book is specifically focused upon design analysis technology for
the class of materials referred to as *engineering thermoplastics*. The
highest selling (by volume) materials classified as engineering thermoplastics are shown in Fig. 2.2 in terms of the total U.S. sales. Since
this book focuses on structural design and since engineering thermoplastics provide an attractive combination of structural properties
and cost, it is logical to focus on these particular polymers. Table 2.1
shows nominal properties of typical unfilled engineering thermoplastics and compares these properties with steel. The lower absolute
mechanical properties of plastic materials must be balanced with opportunities for lower weight (based on specific gravity) and production
efficiencies (such as through the injection molding process).

Crystalline and amorphous polymers

There are also two major thermoplastic material types: crystalline and
amorphous polymers. A *crystalline* polymer is a polymer chain that ex-

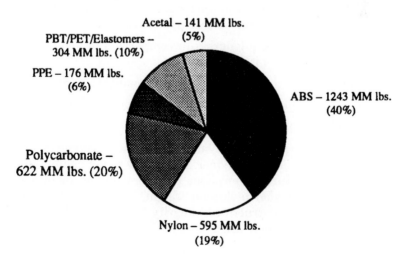

1989 Sales - 3.1B lbs.

Acetal – 141 MM lbs. (5%)

PBT/PET/Elastomers – 304 MM lbs. (10%)

PPE – 176 MM lbs. (6%)

ABS – 1243 MM lbs. (40%)

Polycarbonate – 622 MM lbs. (20%)

Nylon – 595 MM lbs. (19%)

Figure 2.2 U.S. sales of select engineering thermoplastics. (*Source:* Modern
Plastics Encyclopedia.)

TABLE 2.1 Nominal Properties of Typical Unfilled Engineering Thermoplastics with Comparison to Steel

	SI units	English units	Approximate ratio of steel properties to thermoplastic properties
Specific gravity	1.2	1.2	7
Elastic modulus	2 GPa	300 ksi	100
Tensile strength	55 MPa	8 ksi	10
Melting point	200°C	390°F	14
Coefficient of thermal expansion	$7.0 \times 10^{-5} \frac{\text{mm}}{\text{mm}} \big/ °C$	$3.9 \times 10^{-5} \frac{\text{in}}{\text{in}} \big/ °F$	0.15
Thermal conductivity	0.2 W/m · K	$1.4 \dfrac{\text{Btu} \cdot \text{in}}{\text{h} \cdot \text{ft}^2 \cdot °F}$	200

hibits an ordered molecular structure. The term *crystalline* is actually a misnomer since crystalline polymers are actually only semicrystalline in nature. They have regions of ordered molecular structure and also have regions of no order or form (amorphous). Because of their structure, semicrystalline polymers exhibit some unique properties and characteristics. A list of some of these characteristics follows:

- a well-defined melting point
- more resistant to solvents than amorphous thermoplastics
- more shrinkage and more likely to warp due to anisotropic shrinkage during processing than amorphous thermoplastics
- low melt viscosity (long flow lengths) in comparison to amorphous thermoplastics for similar molecular weights
- very little melt strength and, as a result, not as readily blow-moldable
- more temperature-dependent mechanical properties than amorphous thermoplastics

In contrast to semicrystalline polymers, amorphous polymers are composed of randomly oriented polymer chains and do not exhibit any ordered molecular structure. Amorphous polymers rely on increased polymer chain lengths (higher molecular weight) and physical entanglement of those chains for structural integrity. Some general characteristics of amorphous polymers are as follows:

- undergo a second-order transition
- more uniform and quantitatively lower shrinkage during processing as well as greater post-mold dimensional stability

- high melt viscosities
- more susceptibility to chemical attack than semicrystalline polymers
- option of optical transparency

Polymer blends and alloys

As plastics technology has progressed, polymer blends and alloys have become increasingly important material subgroups because they offer unique combinations of properties of each of their parent polymers. Polymer blends fall into three main categories: miscible, immiscible, and partially miscible. Miscible blends consist of a single polymer phase—two or more polymers that are completely soluble in each other. The structure of such a polymer blend is illustrated in Fig. 2.3. The mechanical properties of a miscible blend are usually weighted averages of the properties of the two components. Figure 2.4 illustrates such a material dependence. Polyphenylene ether (PPE) and polystyrene (PS) form miscible blends over the entire composition range of both components. PPE brings the benefits of a high use temperature and flame retardancy; PS adds improved flow and processability, resulting in good properties for extrusion, blow-molding, or injection-molding applications.

Immiscible blends result from mixing two materials with little affinity for each other. The polymer with the smallest volume fraction is usually a poorly dispersed second phase with little adhesion between the phases. Such a blend is illustrated in Fig. 2.3. However, compatibilization improves dispersion and phase adhesion, yielding blends with useful engineering properties. These properties are ideally the best properties of each material, varying as a step function illustrated in Fig. 2.4. The most practiced example of combining two immiscible polymers is for the purpose of impact resistance. Examples of such immiscible blends are high-impact polystyrene (HIPS), some ABS materials,

 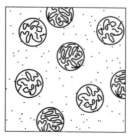 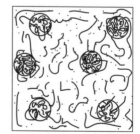

(a) Miscible   (b) Immiscible   (c) Partially miscible blend

Figure 2.3 Possible morphologies of a blend of two polymers.

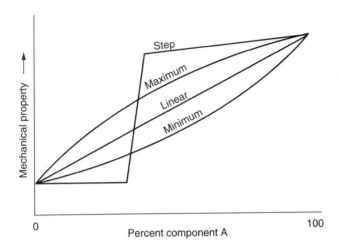

Figure 2.4 Potential responses for the mechanical properties of a polymer blend.

and toughened polyamide [modified rubbery polymers within a polyamide (nylon) matrix]. Another example where chemical compatibilization of two immiscible polymers has lead to a successful new polymer is the blend of PPE with nylon. This blend provides the dimensional stability, high glass transition temperature, and electrical properties of PPE with the improved flow and solvent resistance of nylon.

Some blends are neither completely miscible nor immiscible. These partially miscible blends show limited mutual affinity, but small amounts of one polymer are soluble in the other. The best properties of each blend may be combined, often without the challenge of developing a compatibilization mechanism. Partially miscible blends of crystalline polyesters such as polybutylene terephthalate (PBT) with polycarbonate (PC) yield materials with the dimensional stability and toughness of the PC and the solvent resistance and processability of the crystalline polyester. Impact modification gives these blends good low-temperature toughness.

Reinforcements, fillers, and additives

Short-glass-fiber reinforcement is the most common method used to increase the stiffness and strength of plastics. In addition, creep, temperature, and fatigue resistance typically increase with the addition of glass filler. These increases in mechanical and physical performance can be attributed primarily to the higher mechanical strength of glass versus that of the base polymer. However, the addition of short glass fibers also reduces the strain to failure and thus the ductility and impact resistance of the polymer. Notch sensitivity of glass-filled thermoplastics is generally higher than that of unfilled thermoplastics. And from a processing viewpoint, flow, shrinkage, and coefficient of thermal expansion (CTE) all decrease. The addition of glass to a thermoplastic generally increases the viscosity of the material and makes it more difficult to process. Glass fibers also increase wear on tools and molding machinery. Furthermore, these short glass fibers tend to align themselves along the flow direction of the polymer melt and thus cause anisotropic shrinkage and potential warpage during and after processing. Weld lines, along which flow fronts meet, in glass-filled parts frequently have only the strength of the unfilled resin. Furthermore, the nature of glass fibers in most plastics is to come to the surface of the molded part, thus reducing surface finish and gloss. In addition, the density of a filled thermoplastic increases, as does the typical cost per pound.

Fillers such as minerals, glass spheres, inorganic flakes (e.g., mica), and powders are also frequently added to improve mechanical properties and to control shrinkage. Mechanical properties of the plastic such as stiffness, creep resistance, and temperature resistance increase somewhat, although not as much as with glass filler. As in the case of short glass fibers, ductility and impact resistance decrease while notch sensitivity increases. Furthermore, flow during processing, aesthetics, shrinkage, and CTE all decrease, again as with glass-filled parts. Wear on molding machinery and tools also increases. But unlike thermoplastics filled with short glass fibers, thermoplastics filled with flakes or spheres tend to exhibit isotropic shrinkage properties, thus reducing warpage in complex parts. Typically, fillers of this nature are added to improve specific values, such as electrical properties or flame retardance.

Some fillers can significantly improve arc tracking and volume resistivity. The inherent flame resistance of most fillers increases the flame retardance of thermoplastics when drip or deflection is the criterion for testing. In addition to fillers used to change mechanical properties, other available additives include flame retardants, coloring agents, ultraviolet radiation stabilizers, processing aids, and conductive aids. Most polymers, because they are organic materials, are flammable. Additives such as compounds of chlorine, bromine, phosphorus, or metallic salts can reduce the likelihood that combustion will occur or spread. One of the advantages of thermoplastics is that they can be custom colored using coloring agents, eliminating secondary painting operations. The most common problems arising from extended exposure to ultraviolet (UV) light are color shift in pigmented resins and yellowing of clear and light-colored resins, with possible reduction in mechanical and other properties. There are a number of custom additives that are designed specifically to minimize the effects of UV exposure. Release agents are also commonly added to thermoplastics to enhance processability, specifically in ejection of the part from the tool. Lubricants can reduce the viscosity of the molten plastic and improve forming characteristics. Most polymers, because they are poor conductors (good insulators) of electrical current, build up a charge of static electricity. Some antistatic agents attract moisture from the air, thereby improving the surface conductivity of a part, and conductive fibers can also be added to offer electromagnetic interference (EMI) and radio-frequency interference (RFI) shielding. All additives must be carefully selected because of potential chemical compatibility problems and reduced mechanical properties. The design engineer needs to be aware of opportunities and pitfalls associated with additives.

Engineering Data and Material Selection

Engineering thermoplastics exhibit complex behavior when subjected to constant, increasing, or cyclical mechanical loads. As these materials begin to be used more in load-bearing designs, engineers must be able to predict the structural performance of actual molded parts. However, the necessary engineering design data to do this are usually not available. While standard datasheet properties can be useful for initial material selection, they are inadequate to predict the structural performance of a design. And even when the necessary engineering data exist, they are usually not measured at the same time, strain rate, temperature, or stress as those of a particular application.

Standard datasheets and most computerized databases commonly provide design engineers with three categories of mechanical testing and related properties: flexural testing, heat resistance testing, and impact testing. However, these properties are not directly useful for predicting the structural performance of plastic components.

Flexural testing, for example, involves bending a specimen until it breaks (strength), cycling it (fatigue), or maintaining a constant bending stress (creep). In each of these tests, linear-elastic, time-independent beam equations are used to calculate the bending stress. Unfortunately, because of the time-dependent behavior and nonlinear stress–strain response of thermoplastics, these simple beam equations are often inadequate. Also, in the case of the creep test or a constant-stress fatigue test, constant stress is not actually maintained because of stress redistribution that occurs in the sample.

Additional problems arise because the calculations to arrive at these properties assume isotropic, homogeneous materials. In many situations, tensile and compressive stiffness properties differ. Therefore, a flexural stiffness value represents, at best, an average of this behavior. Since most deformations produce a combination of tension and bending, the average behavior represented by flexural stiffness values can be very misleading.

Flexural strength measurement is further complicated because many unfilled thermoplastics do not visibly fail (i.e., they do not break in two) in bending tests. In addition, typical datasheet values of flexural stiffness and strength are measured at room temperature and a single loading rate. It is well known that the applied strain rate and temperature have a significant effect on the mechanical behavior of thermoplastics.

Similar problems exist with heat resistance properties. The effect of temperature on the mechanical behavior of thermoplastics is most commonly reported as the heat deflection temperature (HDT). An HDT test is performed at a very low constant stress. The temperature is increased until a specific deflection is produced. Such a test, which involves variable temperature and arbitrary stress and deflection, is of no use in predicting the structural performance of a thermoplastic at any given temperature or stress.

For measuring impact resistance there are a number of different test methods in use, such as Izod, Gardner, and Dynatup. The goal of any impact resistance test should be not only to measure the magnitude of energy absorption, but also how performance is affected by changes in temperature. Additionally, strain-rate dependent transitions from ductile to brittle behavior are important. None of the typical measures of impact can be used directly to predict the behavior of a structure under a high rate of loading. Most measurements of impact resistance are also dependent on the geometry of the test specimen. Energy absorption values, however measured, represent many complex processes of elastic and plastic deformation, notch sensitivity, and fracture processes of crack initiation and propagation. Furthermore, there is no method to calculate stress, strain, or strain rate. Finally, the ductile–brittle transition is strain-rate dependent and, therefore, different for each technique.

The design engineer's task is to predict the performance of a design at end-use conditions in terms of both operating temperature and loading (constant, increasing, or cyclic). To do this, two types of information are required: data to perform structural analysis calculations and data to assess performance. In this book, the engineering data requirements for structural design analysis of thermoplastic components will be presented. The shortcomings of datasheet data and standard tests will be overcome with engineering design techniques for thermoplastic components.

Processing Considerations

The design engineer must also be aware of the interaction between processing and design. Almost all design rules and techniques are related to a particular process. The optimum process must be selected and the advantages of the process must be maximized. In this section the processing procedures will be summarized (along with advantages and disadvantages) for injection molding, structural foam molding, extrusion blow molding, compression molding, thermoforming, and extrusion. But first, the influence of processing on material properties and part performance is presented along with general considerations of part and mold design for injection molding. The goal is to provide a general understanding of fabrication processes as they relate to structural engineering design analysis but not to provide detailed part design methods related to specific manufacturing methods.

Even in the simplest amorphous resins, the physical state and local mechanical properties of the injection-molded part material are influenced by the processing conditions that determine the "frozen" state of the material and molecular orientation effects induced by flow. And, because of continuing relaxation processes, the frozen state continues to change; this physical aging results in a continuing change of the mechanical properties well after the part has been molded. In crystalline materials, the degree of crystallization, crystallization gradients throughout the part, and the deformation of spherulites caused by flow again have a marked effect on local properties. Furthermore, in both amorphous and semicrystalline materials, the molding process results in a locked-in residual stress pattern that is controlled by the time–temperature–pressure processing history of the entire part. When such parts are subjected to a temperature history—such as that in a paint–bake temperature cycle or in-service conditions—stress relaxation causes dimensional changes. Because of the complex nature of the influence of the fabrication process on material properties, it is impossible to include these effects in the engineering design process. However, the design engineer should be aware of these effects and allow for them in the design process.

There is a significant amount of experience-based information on part and mold design related to fabrication processes. Although the presentation of these "rules" is beyond the scope of this book, a summary of the type of information that is available from material suppliers, for example, for injection molding includes recommendations such as

- adequate wall thickness for proper flow
- uniform wall thickness for optimum moldability

- minimum wall thickness for minimum cycle time
- adequate draft angle for inside and outside walls to ensure easy part removal
- thin enough ribs to avoid sink marks and voids
- bosses with limited outside and/or inside diameter and wall thickness

All of these recommendations as well as others must be considered along with the structural design techniques presented in the following chapters.

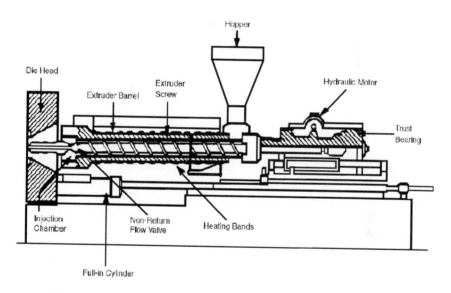

Process Procedure
1. Plastic raw material loaded into hopper
2. Gravity dropped to extruder screw
3. Moved through a heated extruder barrel (material softened and made fluid)
4. Fluid plastic shot into mold
5. Part cools
6. Mold opened and part ejected

Advantages
- Fast cycle times
- High degree of part complexity
- Dimensions and shapes accurately controlled and predicted

Disadvantages
- High pressure process
- Only thin wall sections
- High initial tooling cost
- Poor large-part capabilities

Figure 2.5 Injection molding.

Process Procedure

Same as injection molding except
foaming agent also loaded into hopper
creating bubbles of gas (when heated)
that produce a porous core layer.

Advantages

- High stiffness to weight
  ratio
- Low pressure process
- Large part capability
- Enhanced chemical
  resistance over
  injection molded parts

Disadvantages

- Poor surface
- Longer cycle times
  than injection
- Mix blending required

Note: Foam core can also be achieved by injecting an inert
gas (nitrogen) directly into mold cavity.

Figure 2.6 Structural foam molding.

Processing procedures for various fabrication techniques are summarized in Figs. 2.5 to 2.10, along with the advantages and disadvantages of using these processes. This general information is necessary background for the presentation of structural design techniques related to some of these processes discussed in subsequent chapters. Also, it is useful information for the design engineer who is challenged with the selection of the optimum process for a given application and design. Additional useful criteria for process selection are included in Table 2.2. This process selection table, combined with a resin manufacturer's processing guide, should provide the design engineer with the rudiments of process knowledge. This manufacturing information is an important element of the engineering design process.

Design Engineering

Engineering plastics are now entering markets where their mechanical performance must meet increasingly demanding requirements. And because the marketplace is more competitive, companies cannot afford overly designed parts or lengthy product development cycles. Therefore, design engineers must have design technologies available that allow them to productively create the most cost-effective design with the optimum selection of material and process. In general, the real chal-

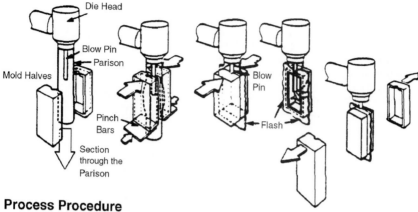

Process Procedure

1. Extruder screw plasticizes material
2. Homogeneous melt formed
3. Parison is formed through die
4. Parison is pinched and partially inflated (pre-blown) in order to help the parison retain its shape
5. Mold closes and parison is inflated
6. Part cools
7. Mold opens, blow pin retracted and part air ejected
8. Mold rises to new parison

Advantage
• Double wall geometry (hollow)
• Low cost tooling
• Large part capability

Disadvantages
• Long cycle time
• Secondary trimming
• High purging start-up cost

Note: With injection blow molding the parison to be blown is formed by injection molding

Figure 2.7 Extrusion blow molding.

lenge to engineers designing with structural plastics is to develop an understanding of not only engineering design techniques but also processing and material information. This process is summarized in Fig. 2.1. The design element involves creating geometry and performing engineering analysis to predict part performance. The material element involves material characterization to provide engineering data. And the processing element includes process and process–design interaction knowledge. The design engineering process involves meeting the end-use requirements with the lowest-cost design, material, and processing combination.

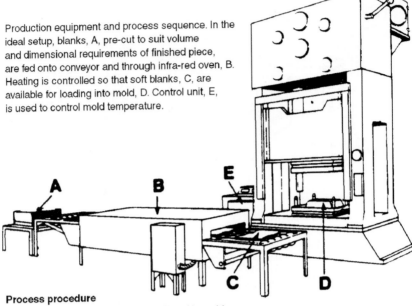

Production equipment and process sequence. In the
ideal setup, blanks, A, pre-cut to suit volume
and dimensional requirements of finished piece,
are fed onto conveyor and through infra-red oven, B.
Heating is controlled so that soft blanks, C, are
available for loading into mold, D. Control unit, E,
is used to control mold temperature.

Process procedure
1. Pre-weighted charge (blank) positioned in mold
2. Mold closed and pressure applied to hot blanks
3. Part removed to cool
4. Flash trimmed

Advantages
- High degree of part density
- Low tendency toward warpage
  and distortion
- Large variation in wall thickness is
  possible
- Excellent property retention after molding

Disadvantages
- High tooling cost (cheaper than injection
  molding tool)
- Secondary trimming

Figure 2.8 Compression molding.

The development of any product requires careful attention to antici-
pated end-use requirements; thus the importance of establishing and
analyzing these requirements cannot be overemphasized. The design
engineering process is keyed to satisfying these end-use requirements.
Specific details needed to establish end-use requirements are included
in Table 2.3.

Basic design rules for thermoplastics can easily be found, for example, in design guides provided by resin suppliers. However, advanced
engineering design techniques are usually required to deal with the

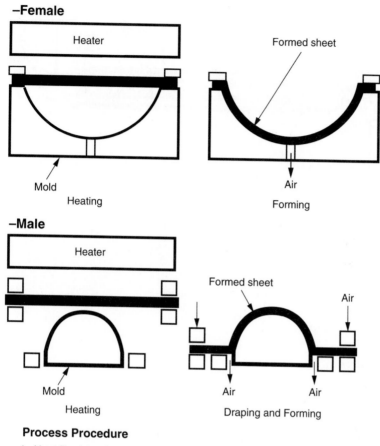

–Female

Heater

Mold

Heating

Formed sheet

Air

Forming

–Male

Heater

Mold

Heating

Formed sheet

Air

Air

Air

Draping and Forming

Process Procedure

1. Heat film (sheet)
2. Position film against die
3. Draw vacuum
4. Allow part to cool
5. Remove part
6. Trim size

Advantages

- Easily automated
- Low tooling cost
- Large-and-small part capabilities
- Low pressure process

Disadvantages

- Longer cycle times
- Low part complexity

Figure 2.9 Thermoforming.

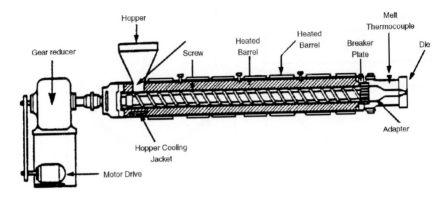

Process Procedure

1. Plastic material drawn from hopper into heated barrel
2. Material is plasticized and mixed as screw moves it toward the die
3. Die sets the shape of the extrusion
4. Plastic leaves die and enters sizing unit to cooling unit
5. Part is cut to required size or coil

Advantages

- Long simple shapes
- Low tooling die cost
- Continuous process
- Small and large part capability

Disadvantages

- Only constant cross section
- Limited part complexity
- High set up cost

Figure 2.10   Extrusion.

TABLE 2.2   Process Selection Criteria

	Injection molding	Structural foam	Compression molding	Blow molding	Thermoforming	Extrusion
Part complexity 1 = most complex	1	1	3	4	4	5
Part size 1 = largest	4	2	1	2	1	1
Tooling costs 1 = least expensive	5	4	5	3	2	1
Secondary operations 1 = least cost	1	3	3	3	3	1
Minimum cost efficient production quantity	10,000	2,500	10,000	2,000	50	50

NOTE: These criteria are for comparison only. Your actual experience will depend upon a number of variables such as design, processing, equipment, and environment, and therefore may vary.

complexities of thermoplastic material behavior, processing, and part geometry. Computer-aided engineering (CAE) techniques for thermoplastics involve the application of computer software and graphics to the engineering analysis of a plastic part and mold. Typically this involves stress analysis to ensure the structural performance of the plastic part and process simulation to optimize the molding process. Overall, computer-aided engineering analysis can lower costs and save time by reducing the elapsed time from conceptual design to actual part production. This increased productivity results from efficient computer techniques and the increased part quality and moldability ensured by analysis-based design iterations. The costly and time-consuming redesign necessary after a prototype part fails structurally or does not fill the mold properly can be avoided with this analysis. Also, there are gains made in structural performance, improved part quality, and reduced manufacturing costs through reduced cycle times.

TABLE 2.3   End-Use Requirements

Structural Requirements
Magnitude of applied load and/or deflection
Rate of loading
Duration of constant load and/or deflection
Number of cycles
Maximum allowable deflection
Environmental Requirements
Temperature extremes
Exposure to chemicals and water
Flammability
Electrical Considerations
Voltage requirements
Tracking requirements
Insulation requirements
Appearance Considerations
Style, shape, color, surface finish
Transparency
Codes and Specifications
Food and Drug Administration (FDA)
Underwriters Laboratory (UL®)
National Sanitation Foundation (NSF)
Society of Automotive Engineers (SAE)
American Society for Testing and Materials (ASTM)
American National Standards Institute (ANSI)
International Standards Organization (ISO)

Closure

The focus of this book is the structural engineering design analysis of thermoplastic components. Engineering analysis is an integral and fundamental part of the design engineering process (Fig. 2.1). Preliminary and final detailed design (involving the creation of part geometry) and prototype fabrication and testing are also important steps in the design engineering process. However, the emphasis of this book is on the engineering analysis phase: analysis tools and techniques, engineering data, and process–material–design interactions. At this point, some general material and manufacturing aspects of the design engineering process have been covered. Now, with this background, the specifics of structural design analysis of thermoplastic components can be treated in the following chapters.

References

1. *Modern Plastics Encyclopedia,* McGraw-Hill, New York; annual publication.
2. D. V. Rosato, D. P. DiMattia and D. V. Rosato, *Designing with Plastics and Composites: A Handbook,* Van Nostrand Reinhold, New York, 1991.
3. R. J. Crawford, *Plastics Engineering,* Pergamon, New York, 1987.
4. A. W. Birley, B. Haworth, and J. Batchelor, *Physics of Plastics-Processing, Properties and Materials Engineering,* Hanser, Munich, Germany, 1992.
5. Joel Frados, *The Story of the Plastics Industry,* Society of the Plastics Industry, New York, 1977, pp. 37–50.
6. Cyril Dostal, Ed., *Engineered Materials Handbook,* American Society of Materials, Vol. 2, *Engineering Plastics,* Materials Park, OH, 1988, pp. 277–399.

Finite-Element Analysis Techniques and Nonlinear Issues

Once the functional requirements of an engineering structure have been identified and concepts for its general shape have been defined, the engineering design process can be focused upon the goal of defining the most effective geometry to ensure performance. There are usually several competing measures of effectiveness for this, including cost, weight, style, and ease of assembly that must be assessed and compared. There are also different approaches to achieving an engineering solution. For example, part concepts can be built, tested, altered, and then retested until adequate measures of effectiveness are reached. Alternatively, engineering analysis can be used to provide answers to some of the questions of design effectiveness before fabrication. Although component tests still usually are the final measure of success or failure, analysis is increasingly being applied to reduce the design time and lower the cost required to achieve an effective part.

Engineering analysis in support of design can take different forms. Sometimes simple formulas tabulated in engineering handbooks are quite adequate for defining geometric dimensions that will fulfill performance requirements. In other situations these classical approximations may be less straightforward to apply, either because of complex part geometry or significant nonlinear physical effects. Over the past 35 years, one of the most significant developments in analytical methods, directed at overcoming these issues of geometric complexity and nonlinearity, has been the finite-element analysis method (FEA or FEM). This method has been used successfully in a wide range of industries, including aerospace, automotive, and consumer products, to name a few.

Just as with other engineering materials, the finite-element method has become an extremely important tool for analysis of components made of thermoplastics. Part fabrication processes such as injection molding provide a means of creating geometrically complex parts, such as the industrial pallet shown in Fig. 3.1—parts that make simple handbook formulas difficult to apply. Furthermore, the low material stiffness and yield strengths of thermoplastics create situations where nonlinear behavior is often commonplace. Figure 3.2 illustrates an example of the large displacements and inelastic material behavior that can occur in plastic parts. The necessity of treating these nonlinear conditions routinely is another reason why this method has seen such heavy use in the analysis of plastic parts.

Throughout this book, repeated use will be made of the finite-element method to identify significant issues and provide evidence of the accuracy with which engineering analysis of plastic parts can be conducted. Although it is not within the scope of this text to provide a detailed description of the underlying theory of this analysis approach, some general level of understanding of this analysis method will be useful. A general description of the basic theory behind finite-element analysis will be followed by a discussion of element types commonly available for application. The generic procedure for applying finite elements will also be summarized in terms of geometry definition and mesh creation, boundary and load conditions, material properties, and interpretation of results. The purpose of the chapter is to provide a

Figure 3.1 Thermoplastic industrial pallet.

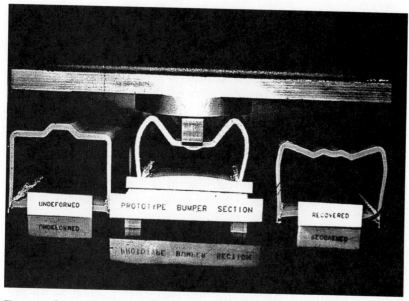

Figure 3.2 Large displacements and inelastic material behavior in a thermoplastic box section.

simplified basis for understanding the application of this technique during the rest of the book. For a more detailed understanding of FEA, the reader is referred to any of the many books on this subject.[1-6]

Since ability to account for nonlinear behavior is one reason that finite-element techniques are applied, a basic understanding of some of the nonlinear structural behavior that is significant for plastics will also be very useful during the course of reading this text. As mentioned previously, the low material stiffness of plastics often leads to situations where nonlinear behavior is much more important to the understanding and design analysis of these materials than with other engineering materials. Just as in the case of the finite-element analysis method, a complete treatment of nonlinear behavior will not be attempted in this text. However, there are several basic nonlinear concepts that are very significant to understanding the behavior of thermoplastic materials and components that will be presented.

Structural Finite-Element Analysis: Theoretical Basis

One of the foundations of structural mechanics is Hooke's discovery that, for many structures, the relationship between force and deflection could be practically approximated as linear in nature. Using a simple spring as an example of a structure, its stiffness is simply the force divided by the deflection, as long as the spring remains linearly elastic and does not yield or break. If one applies twice the force, the spring will deflect twice as much. A bar of cross-sectional area A and length L made of a material with Young's modulus E is one specific and simple example of a structural spring. For such a bar, the structural stiffness can be written as

$$S = \frac{P}{\delta} = \frac{EA}{L} \tag{3.1}$$

As can be seen in Eq. (3.1), the structural stiffness is dependent on both material properties such as E and geometric properties such as A and L.

Although the stiffness of a bar is easily defined in terms of its geometric properties, most engineering components exhibit far more complex geometry, making accurate structural analyses much more difficult. In mid-1950s, American and European aeronautical engineers independently developed the finite-element method as an approach to analyzing such structures.[7] Using this approach, a structure can be idealized as being composed of many small, discrete pieces called finite elements. These engineers extended Hooke's basic idea of linear springs into a general approach capable of analyzing extremely large structures. As a direct result of the concept of breaking the complex structure down into smaller, simpler pieces, the problem now became characterized by large numbers of simultaneous equations. The advent of the first generation of computers soon made the solution of these equations straightforward.

In the 1960s, the finite-element method began to be applied more broadly, including fields of engineering science, such as fluid mechanics, heat transfer, electromagnetics, wave propagation, and other field problems. Applied mathematicians developed criteria that ensured convergence of these numerical solutions as smaller elements (or pieces) were used. As linear solutions became routine, techniques for application of the FEA approach to nonlinear problems received increased attention. Today, solutions to structural component problems with complex geometries and nonlinearities, encompassing both material and geometric behavior, have become increasingly commonplace.

FEA simulation has become a very important tool in engineering part design in general and plastic parts offer no exception. When used intelligently, this method can provide a wide range of advantages, including

- improved performance and quality of the final plastic product,

- optimum use of materials,

- weight savings,
- improved probability of successful performance of prototype parts,
- faster time to market and therefore a competitive edge, and
- reduced development, production, and part costs.

Fundamentals

The basic theory of FEA is to reduce a large, complex structure into a network of small, simple geometric elements, such as beams, two-dimensional (2-D) elements, shells, or solids (3-D elements). Within any of these elements, relatively simple equations can be used to describe the measures of deformation, e.g., stress, strain, and displacement. The behavior of the entire structure is calculated by combining the element equations into a large set of simultaneous equations representing the behavior of the structure.

In its current form, finite-element analysis is not only appropriate for linear static problems, but can also treat dynamic and nonlinear problems as well. In the simplified overview that follows, only linear static situations will be discussed. Linear analysis applies to problems where stress and strain are related in a linear fashion and deflections are small. In contrast, nonlinear analysis is required when material constitutive relations or boundary conditions become nonlinear in nature or when strains and displacements become large.

In order to gain a general understanding of FEA theory, consider its application to a simple two-dimensional problem. In Fig. 3.3 a general two-dimensional structure has been divided into a number of 2-D, triangular finite elements and associated nodes. A finite-element model can be thought of as a system of springs. When a load is applied to the structure, all of the elements must deform in a fashion that guarantees equilibrium of forces between the elements. In addition, the deformation of the modeled structure must remain compatible. This later requirement must be fulfilled in order to ensure that discontinuities in displacement do not develop at element boundaries. Let us consider the development of these equations for a structure modeled with 2-D triangular elements like the ones in Fig. 3.3.

The first step in developing these equations is to establish the expression for element stiffness, relating forces and displacements at the nodes of an element. The sequence in this process is as follows:

1. Assume an approximate displacement function for the element. This function is defined in terms of the displacements at the nodes of the element and should ensure compatibility of displacements with neighboring elements along its entire boundary.

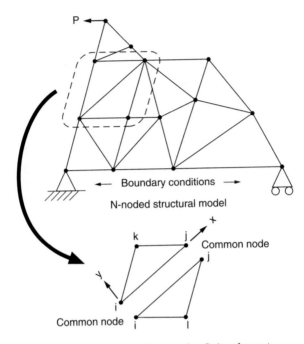

Figure 3.3 Two-dimensional triangular finite-element model of a hypothetical structure.

2. Apply the kinematic equations defining strain in terms of the approximate displacement functions.

3. Use the *constitutive* relationship appropriate for the material to determine stresses in terms of strains.

4. Develop *equilibrium* equations relating internal element nodal forces to externally applied nodal forces.

Displacement function

The displacement function characterizes the displacements within an element as a function of space. The choice of displacement function affects the accuracy of the element in approximating actual displacement, strain, and stress behavior over the volume of the element. Since strains are first derivatives of displacements, a linear displacement function leads to the approximation of constant strains and stresses within the element. Similarly, a quadratic displacement function simulates linear stress and strain fields within an element. For the three-node, triangular element shown in Fig. 3.3, we will designate the x axis to lie along one edge of the triangle. With displacements (U_i, V_i) in two coordinate directions (x,y) at each node (i) there are a total of six nodal displacements (degrees of freedom) in terms of which the deformation field for the element can be defined. In order to define the displacements within the element in terms of these six nodal displacements, functions with a total of six coefficients are required. A natural set of choices for this element is

$$u = a + bx + cy \qquad (3.2)$$

$$v = d + ex + fy \qquad (3.3)$$

where a, b, c, d, e, and f are unknown constants.

Element displacements in terms of nodal displacements

Using Eqs. (3.2) and (3.3) to evaluate the displacements at each node (i),

$$U_i = a \qquad (3.4)$$

$$U_j = a + bX_j \qquad (3.5)$$

$$U_k = a + bX_k + cY_k \qquad (3.6)$$

where X_j, X_k, and Y_k define known coordinate locations. A similar set of equations can be written to define the coefficients d, e, and f in terms of the nodal y displacements. Using matrix manipulation it is possible to represent the u and v displacements within an element in terms of nodal displacements as

$$\begin{Bmatrix} u(x, y) \\ v(x, y) \end{Bmatrix} = [N] \begin{Bmatrix} U_i \\ V_i \\ U_j \\ V_j \\ U_k \\ V_k \end{Bmatrix} = [N]\{\delta\} \qquad (3.7)$$

where the $[N]$ is the "shape function" matrix that can be formed by using Eqs. (3.2) to (3.6), and $\{\delta\}$ is the vector of nodal displacements.

Strains as a function of nodal displacements

The two-dimensional definitions of strain in terms of displacement

$$\varepsilon_x = \frac{\partial u}{\partial x} \qquad (3.8a)$$

$$\varepsilon_y = \frac{\partial v}{\partial y} \tag{3.8b}$$

$$\gamma_{xy} = \frac{\partial u}{\partial y} + \frac{\partial v}{\partial x} \tag{3.8c}$$

are then used to calculate the strains within the element in terms of its nodal displacements. Using matrix notation again, this relationship can be expressed as

$$\{\varepsilon\} = \begin{Bmatrix} \varepsilon_x \\ \varepsilon_y \\ \gamma_{xy} \end{Bmatrix} = [B]\{\delta\} \tag{3.9}$$

where $[B]$ is a matrix that can be defined in terms of derivatives of the shape function elements using Eq. (3.8).

Stresses in terms of strains

In order to relate stresses to strains, a material constitutive model is necessary. For simple linear elasticity, the plane-stress constitutive relations are

$$\sigma_x = \frac{E}{1 - v^2}(\varepsilon_x + v\varepsilon_y) \tag{3.10a}$$

$$\sigma_y = \frac{E}{1 - v^2}(\varepsilon_y + v\varepsilon_x) \tag{3.10b}$$

$$\tau_{xy} = \frac{E}{2(1 + v)}\gamma_{xy} \tag{3.10c}$$

where E is the elastic modulus and v is Poisson's ratio. Using matrix notation,

$$\begin{Bmatrix} \sigma_x \\ \sigma_y \\ \tau_{xy} \end{Bmatrix} = [D] \begin{Bmatrix} \varepsilon_x \\ \varepsilon_y \\ \gamma_{xy} \end{Bmatrix} = [D][B]\{\delta\} \tag{3.11}$$

where $[D]$ is the material matrix formed using Eq. (3.10).

Nodal forces in terms of displacements

Load is transmitted from one finite element to another through forces at the node points of the elements, which can be represented as $\{F\}$. These nodal forces in the two coordinate directions are related to the nodal displacements through a set of element equilibrium equations. These equilibrium equations can be defined by equating the external work accomplished by the nodal forces when subjected to an arbitrary set of virtual nodal displacements, $d\{\delta\}$, to the internal energy stored in the element's volume as its stress field is subjected to the virtual strain field resulting from the same virtual nodal displacements. This relationship can be expressed as

$$(d\{\delta\})^T\{F\} = \int_{\text{Vol}} d\{\varepsilon\}^T\{\sigma\}\, d\text{Vol} \qquad (3.12)$$

Since the virtual strains can be related to the virtual nodal displacements as

$$d\{\varepsilon\} = [B]\, d\{\delta\} \qquad (3.13)$$

the element equilibrium Eq. (3.12) takes the form

$$d\{\delta\}^T\{F\} = d\{\delta\}^T\int_{\text{Vol}} [B]^T[D][B]\{\delta\}\, d\text{Vol} \qquad (3.14)$$

Equation (3.14) now takes the form of a relationship between the nodal forces $\{F\}$ and the nodal displacements $\{\delta\}$,

$$\{F\} = [K]_e\{\delta\}, \qquad (3.15)$$

where $[K]_e$ is the element stiffness matrix defined as:

$$[K]_e = \int_{\text{Vol}} [B]^T[D][B]\, d\text{Vol} \qquad (3.16)$$

Global equilibrium

Equation (3.15) establishes the relationship between the nodal displacements of an element and the corresponding nodal forces. When individual elements are joined at common nodes to model a structure such as that shown in Fig. 3.3, global equilibrium must be ensured at each node. This requirement means that the summation of the forces associated with all the elements attached to that node must be equal in magnitude and opposite in direction to the externally applied force at that node.

To construct these equations, individual element stiffnesses are assembled using matrix algebra techniques into a global stiffness matrix representing the stiffness of the entire structure. Since different elements in the model will share common nodes (as shown in Fig. 3.3), individual components of element stiffness matrices are added to form the global set of matrix equations. This global set of equations relates all the nodal degrees of freedom in the structure to the externally applied nodal forces. If the externally applied forces are known, a solution for the nodal degrees of freedom can be obtained using linear algebra once the required boundary equations are applied. When the displacements of all the nodes are known, the state of deformation of each element is also defined. Thus, the state of stress and strain within each element can be calculated using Eqs. (3.9) and (3.11). However, since equilibrium is only guaranteed at a finite number of nodal points in the structure, the finite-element method is a numerical approximation rather than an exact solution. The accuracy of the approximation will depend on the number of nodes and elements in the structure.

Types of Finite Elements

Although the previous description of the equations associated with the equilibrium of a finite element was described for a two-dimensional element, there are many other types of elements. The choice of the type of element to use for a particular problem is often a trade-off between reality and simplicity. Elements can be categorized as one-, two-, and three-dimensional solid elements and beam, plate, and shell elements. The complexity of the analysis and the amount of engineering and computer time required increase significantly when moving from 1-D to 2-D to 3-D analysis. Many real part geometries and loadings are certainly 3-D in nature. When 2-D or 1-D elements are used, assumptions must be made relative to the distribution of stress and strain in the other directions. Since plastic parts tend to have thin walls relative to their overall dimensions, plate- or shell-type elements are often most suitable. Plate or shell analysis can treat the geometric complexity adequately and offer the flexibility to change the wall thickness of the model easily during engineering parameter studies, whereas with fully 3-D analysis, a thickness change requires the nodes of the finite-element mesh to be moved, which is usually a more time-consuming process. However, plate and shell elements are complex and vary widely in their formulations. Arguments continue to persist with respect to the relative ability of the different commercially available elements to accurately predict part performance. In the following section, different finite-element types are briefly discussed. For a more detailed understanding of element types, the reader is referred to books on FEA[1-6] and to various finite-element codes such as those in Refs. 8 to 11.

One-dimensional elements

Bar or truss elements illustrated in Fig. 3.4 are simple, one-dimensional elements. Their length is calculated from the nodal positions defining the bar ends. The cross-sectional area is defined by the modeler. These elements are really nothing more than simple spring elements.

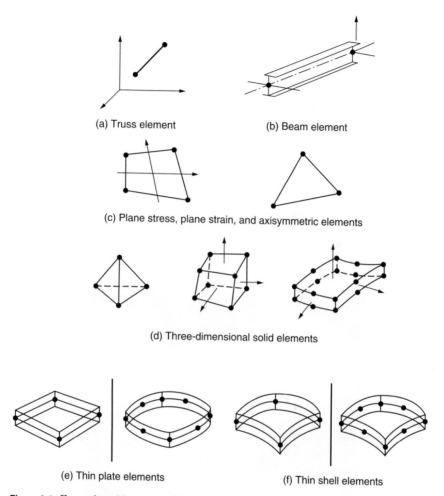

(a) Truss element   (b) Beam element

(c) Plane stress, plane strain, and axisymmetric elements

(d) Three-dimensional solid elements

(e) Thin plate elements   (f) Thin shell elements

Figure 3.4 Examples of finite elements.

Two-dimensional elements

Two-dimensional elements, also illustrated in Fig. 3.4, include plane-stress, plane-strain, and axisymmetric elements. The plane-stress assumption (that stress in the thickness direction is zero) is used when the component's deformation is independent of the dimension perpendicular to the plane of description and its thickness in that direction is small. Plane stress is an appropriate assumption for the thin "snap-fit" beam shown in Fig. 3.5a. The plane-strain assumption (that strain in the thickness direction is zero) is used when the component is thick relative to the planar dimensions. An example of a structure where

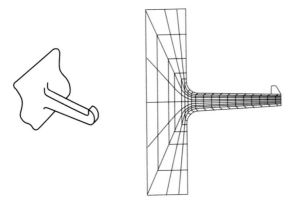

Figure 3.5a Thin snap-fit arm modeled with plane-stress elements.

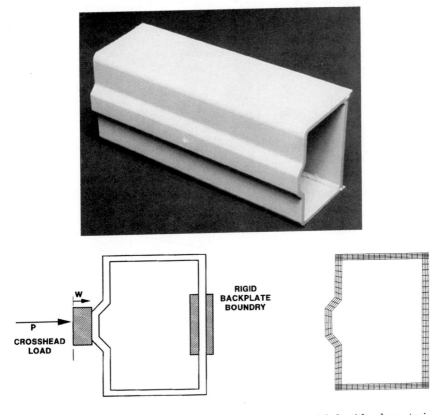

Figure 3.5b Long, two-dimensionally loaded plastic box modeled with plane-strain elements.

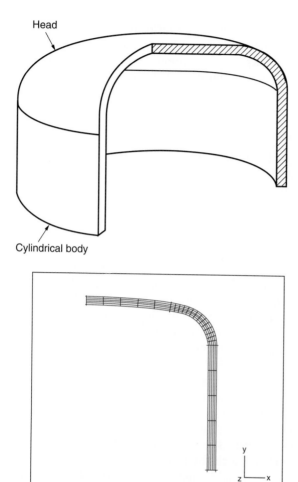

Head

Cylindrical body

Figure 3.5c Plastic bottle modeled with axisymmetric elements.

plane strain is a reasonable assumption is the long box section shown in Fig. 3.5*b*. In both cases of plane stress and strain, the forces acting on the elements as well as the stress and strain response are planar. Hence, there can be no variation of load, geometry, or boundary conditions in the thickness direction. In the axisymmetric element, the stresses, strains, and loads do not vary in the circumferential direction. These elements are often quite useful because they account for fully three-dimensional behavior. Figure 3.5*c* schematically illustrates one end of an internally pressurized container and the axisymmetric finite-element mesh used to model it. When applying these elements, however, care must be exerted to ensure that the loads and expected response show no variation in the circumferential direction. This is particularly important for nonlinear problems, since some nonlinear response to axisymmetric loads may be nonaxisymmetric in nature. An axisymmetric element will not account for this response, even if it is formulated to include nonlinearities.

Three-dimensional elements

Three-dimensional elements are typically either tetrahedrons or hexahedrons, as shown in Fig. 3.4. One of the most common is the rectangular hexahedron with eight nodes, one at each corner. This is analogous to the three-node triangular element, both having linear displacement fields within the element. Normally it is not necessary to model plastic parts with 3-D elements because of the thin-walled nature common to their structures. However, such elements will be used in this text to study local stress distributions around geometric details like notches.

In the absence of the availability of beam, plate, and shell elements, three-dimensional elements can sometimes be applied to the analysis of thin structures using one element through the thickness. Many commercial finite-element codes offer 3-D elements with a choice for the number of nodes used to describe the element geometry. For three-dimensional brick elements, this choice of nodes generally ranges from 8 to 20 or 21. The eight-node elements offer linear displacement functions within the element. The nodes in excess of eight are located at the midsides of the line segments connecting the corner nodes and offering quadratic variations in displacement. A three-dimensional element often used to model thin structures is the 16-node brick. This element includes four midside nodes on both the top and bottom surfaces of the thin-walled structure, but no midside nodes through the thickness. As a result, this element enforces a linear distribution of in-plane displacements through the thickness, commonly associated with beam, plate, and shell theory. In addition, the eight nodes on the top and bottom surfaces allow for quadratic variation of the displacements over the in-plane dimensions. This is important since the load-induced curvature of a thin-walled structure is related to the second derivative of the transverse displacement with respect to in-plane coordinates. As a result, the quadratic variation in the plane of these elements provides for constant curvature within a single element. The bending nature of a thin-walled structure can then be modeled with just one of these elements through the thickness of the part.

Beam, plate, and shell elements

Beam, plate, and shell elements are unique in that they are formulated specifically to model general three-dimensional structural components

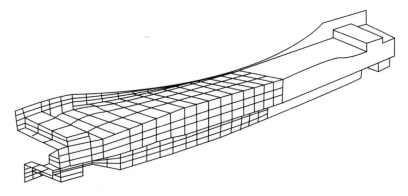

Figure 3.6 Automobile bumper modeled with thin-shell elements.

that are thin relative to their other dimensions. They differ from the 2-D and 3-D finite elements previously described in two major respects. First, all the nodes describing the geometry of these elements are at the midsurface of the component. The thickness of the element is usually specified as either a nodal or element parameter, not unlike the cross-sectional area of the bar elements previously discussed. Second, the degrees of freedom associated with the nodes of these elements now include rotations as well as translations.

As emphasized previously, plate and shell elements are particularly useful in modeling plastic components such as the automobile bumper shown in Fig. 3.6 because of the thin-walled nature of such parts. Formulations for these elements vary, and close attention should be paid to the specific recommendations made relative to the application of these elements by the commercial software companies that provide them. Many of these elements are formulated under the assumption that transverse shear deformation within the element can be neglected. Normally, this is a valid approximation. However, if the beam, plate, or shell is relatively thick or is a sandwich construction with a very low modulus core material, then shear deformation may be significant. There are beam, plate, and shell elements that are specifically formulated to account for shear deformation when it is significant, and users should take note of the presence or absence of this capability when using these elements.

Finite-Element Analysis Procedure

Geometry creation

The first step in the finite-element analysis procedure is to model the part geometry. There are many ways to define geometry, ranging from two-dimensional drawings to three-dimensional computer-aided design (CAD). Computer-aided drafting permits easy generation and editing of two-dimensional geometry. In general, this process involves placing lines, rectangles, arcs, circles, and other basic geometric shapes on a display screen and then moving, rotating, and scaling these shapes to define a part outline. Often, there is a need to describe a part in three dimensions so that it can be more easily understood and converted to a discretized finite-element definition. Wireframe modeling is the simplest approach to graphical display of three-dimensional shapes by definition of part outlines and intersections of surfaces. Unfortunately, these models can be ambiguous and difficult to visualize. Surface modeling goes one step beyond wireframes by describing the individual surfaces of the model, analogous to stretching a thin fabric over the wireframe model. Solid models provide the most unambiguous description of part geometry by mathematically describing the interior and exterior of the part.

No matter how the geometry is created, it must eventually be described discretely in terms of nodal points and elements in order to apply finite-element analysis. For complex parts, this process is usually accomplished by using an automated finite-element mesh generator to represent a part discretely in terms of nodes and elements. In spite of the automated nature of this process, there is often a need to apply engineering analysis judgment. For example, if a shell finite-element model is applied, the engineer must be aware that certain concentrations such as the corners of a box may not be adequately modeled locally with the shell analysis. This is a situation that would require a three-dimensional analysis to define local stresses accurately if that were necessary. Other geometric features such as bosses may sometimes be ignored. Overall, a significant amount of engineering judgment is required to produce an effective geometric representation of a complex part.

Mesh creation and element selection

Once the overall geometry has been defined it must be divided into elements that are connected to one another at the nodal points. This division of the geometry into a set of elements is referred to as a mesh. Engineering judgment is required to select an appropriate element type, as discussed in the preceding section. Also, engineering judgment is required to determine the mesh density and the number and size of the elements. Coarser meshes result in faster solution times but also limit the accuracy of the analysis. Higher mesh densities should be created in regions where large stress gradients are expected. Automated meshing routines are available where the user can specify the mesh density.

Boundary and loading conditions

Boundary conditions on a structure appear as applied displacements at points of support. For static problems, the stiffness matrix associated with the linear equations of equilibrium for the complete structure will be singular, and therefore uninvertible, unless all rigid body motion is prohibited. As a result, a fundamental requirement for solution of the linear equations governing a problem is that the structure must be prevented from freely translating or rotating in space. Rigid body motion is eliminated through the application of boundary conditions requiring zero displacements and/or rotations at nodes. Additional displacement boundary conditions can also be applied to the structure to model the actual structural support system. It is not necessary to restrain all of the displacements and rotations at a node. Selective application of boundary conditions can be used to model sliding of a component over a surface, for example.

Loads may be applied to a model either in the form of applied forces or displacements. Concentrated forces can only be applied at the node locations of the elements. Distributed loads (pressures) and body forces can also be applied to finite-element surfaces and volumes, respectively. These loads are usually internally translated to equivalent nodal forces within the finite-element code.

Material properties

In addition to the geometric detail of the component and the applied loads, the material (constitutive) relationship between stress and strain must also be defined. For simple isotropic, linear-elastic stress analysis, only the material elastic modulus and Poisson's ratio need be provided. In some cases, more detailed constitutive models may be desirable. For example, for highly loaded parts, elastic–plastic behavior may be included. In some cases, nonlinear stress–strain relationships may be represented through a multilinear approximation of the curve. Some material models allow for the definition of strain-rate or time-dependence of a yield condition. Such material models may be useful in modeling impact events or creep situations, respectively. It must be emphasized that increased capability in modeling material behavior means in general that more material data must be available. In many cases, such as time-dependent material models, for example, measurements to obtain such data are nonstandard in nature.

In general, material properties represent the fundamental measurements that relate the performance of a material to the performance of a geometrically complex structural component. It should be emphasized that they play two roles in structural analysis. First, the material properties that define the deformation behavior of a material are used, within the framework of the material's constitutive model, to relate stress to strain in the finite-element method. Second, the material properties that define failure limits are used to interpret the results of an analysis in terms of likelihood of failure. Both of these types of properties may be functions of time, rate, temperature, or other variables. Two things must be kept in mind. First of all, real material properties are not dependent upon geometry. Second, the property is only useful in the general engineering sense if it is associated with a methodology of applying it to general geometries. Many tests are carried out on materials as functions of rate and temperature to provide comparative performance values. However, in many cases these measurements do not represent true material properties because they are uniquely associated with the test geometry used to make the measurement and there is no methodology available to generalize the measurement for use in complex geometries. Within this text, we will focus predominantly upon true material properties and tests while identifying measurements that are only useful to compare material behavior in one particular geometry.

Interpretation of results

After the analyst has defined geometry, element, and node discretization, boundary conditions, loads, and material constitutive relationships, the finite-element code can assemble the equilibrium equations governing the structure. These equations can vary in number from hundreds to thousands for typical problems. The finite-element code solves this system of equations. As a result of this solution, a massive amount of information is computed—displacements of all nodes and stresses and strains in all elements (six stress components for a 3-D analysis). Fortunately, this information can be displayed with advanced graphics techniques as constant-stress contours or with a color-coded representation of the particular stress range of interest. These results can then be assessed in terms of engineering performance requirements. In order to judge whether failure will occur, material data defining failure limits in terms of stress or strain are generally required.

Nonlinear Issues in Analysis

In the previous discussion of the finite-element technique, all of the examples presented were developed under the assumption of linearity. However, it should be emphasized that the finite-element technique can also be applied to analyze nonlinear structural behavior. There are several different types of nonlinear behavior that are important to plastics. The most obvious nonlinearity associated with plastics is the nonlinear constitutive relationship relating stress to strain for some plastics. However, some nonlinearities are geometric in nature and are associated with the strain-displacement equations. A third nonlinearity may be associated with the application of load. In some situations both the magnitude and location of load application depend upon the structure's deformation and cannot be explicitly defined beforehand. In such cases, the contact loads must be calculated as part of the solution process, thus making the problem nonlinear. Modern finite-element techniques can be used to handle all of these problems. In the sections that follow, the fundamental physical and mathematical nature of some of these nonlinearities will be discussed. In many cases the physical properties of thermoplastics make nonlinear response a more important topic than it may be for other materials. Wherever possible, the underlying reasons for this relative importance will be highlighted and discussed.

Nonlinearity due to material behavior

Some nonlinearities encountered in engineering analysis are the result of material behavior. Although the first approximation of constitutive behavior is usually a linear relationship between stress and strain, many common engineering thermoplastics exhibit a very nonlinear stress–strain relationship as can be seen from the material data for

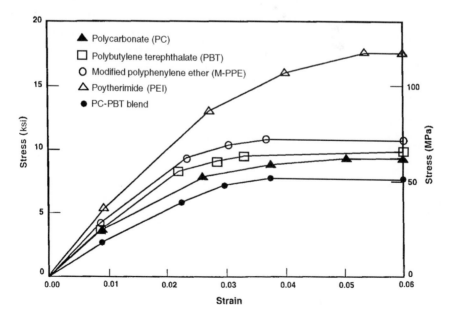

Figure 3.7 Stress–strain relationships for typical engineering thermoplastics.

several thermoplastics in Fig. 3.7. Many FEA analysis codes offer either nonlinear elastic or elastoplastic constitutive models that are useful in accounting for this common thermoplastic behavior.

In order to define the nonlinear stress–strain relationship, a finite set of stress–strain pairs are generally provided by the analyst as illustrated by the symbols in Fig. 3.7. The FEA software uses these pairs to interpolate all other stress–strain values as required during the analysis. Such a solution process is generally iterative in nature and requires more time than required in linear analysis to reach a converged solution. In addition, it is often necessary to increase the load incrementally in such a nonlinear problem for the numerical solution to be efficient. Although this also increases the computer time required for solution, modern computers are capable of providing these solutions routinely and rapidly.

As an example of such a solution, consider the response of a "snap-fit" beam to an end load as shown in Fig. 3.8.[12] The model shown in this figure is made up of 2-D plane stress elements. A number of elements are used through the depth of the beam in order to capture accurately the nonlinear stress distribution encountered as a function of that dimension. For this analysis, a multilinear representation of the stress–strain curve describing polybutylene terephthalate (PBT) is used as illustrated in Fig. 3.7. As the end displacement of the beam increases, the maximum strain on the top and bottom of the beam increases linearly as shown in Fig. 3.9. The maximum stress, however, increases in a nonlinear fashion, which is also shown in Fig. 3.9, and is governed by the material's stress–strain curve. As a consequence of this behavior, for large displacements the force required to increase the beam's displacement incrementally is less than that predicted by a linear elastic analysis, as is also shown in the figure.

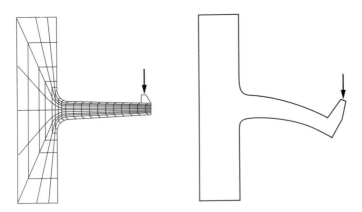

Figure 3.8 End-loaded snap-fit beam.

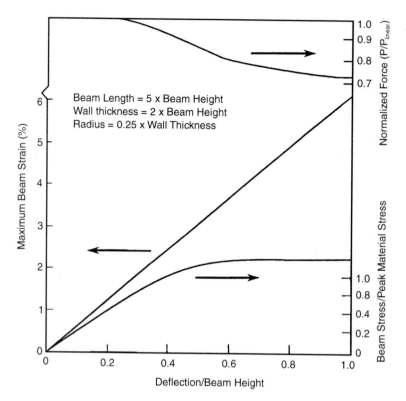

Figure 3.9 Strain, stress, and force as a function of the end displacement of a snap-fit beam.

Nonlinearity due to large displacements in thin plastic components

As previously emphasized, the thin-walled nature of most plastic parts means that the simplified continuum theories of beams, plates, and shells are often effectively used for analysis. An assumption that is regularly made when the customary engineering equations are applied to solve problems of this nature is that the rotations describing the continuum deformation are small (often referred to as the small-displacement assumption). However, the elastic moduli of polymers are routinely as much as two orders of magnitude less than those of metals. Furthermore, plastics will undergo as much as an order of magnitude more strain before incurring damage. These physical phenomena related to plastics can often result in larger rotations and displacements in thin plastic structures than an engineer may be accustomed to seeing with metals.

To develop an understanding of the significance and physical behavior associated with moderately large rotations in thin plastic structures, the typical behavior of beams and plates will be examined as the displacements and rotations imposed on these structures grow. It will be shown that this geometric nonlinearity can progressively increase or decrease the stiffness of a plastic part, depending upon the details of the structure and the applied load. In general, when lateral displacement (displacement perpendicular to the midsurface) of a thin-walled structure takes place in the presence of a tensile, in-plane strain field, large rotations can lead to an increase in stiffness with respect to linear theory. On the other hand, if lateral displacement occurs in the presence of a compressive, in-plane strain field, a decrease in structural stiffness may occur.

The range of applicability of small-displacement linear solutions in terms of displacement and rotation size will also be discussed. For beams, which are one-dimensional in nature, a nonlinear contribution to stiffness depends on the in-plane boundary conditions at the beam ends. As will be shown, there are boundary conditions for which no additional nonlinear stiffness is generated, and as a consequence small-displacement theory is accurate. However, in contrast to the one-dimensional beam, a laterally loaded plate with support on all four edges exhibits nonlinear stiffening for all possible in-plane boundary conditions. The exact conditions at the boundary will affect the amount of added stiffness, but it will always be present to some extent. It should be emphasized that the nonlinear effects upon stiffness to be discussed here are a consequence of purely geometric considerations. However, the importance of these effects for plastic structures in comparison to those made of metal is directly related to the large yield strains associated with plastic materials. Examples illustrating this statement will be provided as part of the following discussion.

Nonlinearity due to large rotations in beams and plates

Consider the simply supported beam under a concentrated, centrally located load, shown in Fig. 3.10. Using the standard Euler–Bernoulli beam theory, the strain in the x direction of the beam is described in terms of the u (displacement in the x direction) and w (displacement in the z direction) displacements of the neutral axis of the beam. For the problem considered here, it will be assumed that the cross section of the beam is a simple rectangle, as shown in Fig. 3.10, thus making the neutral axis pass through the center of the cross section. If the fully nonlinear, one-dimensional strain–displacement relations of continuum mechanics are used, along with the Euler–Bernoulli assumption

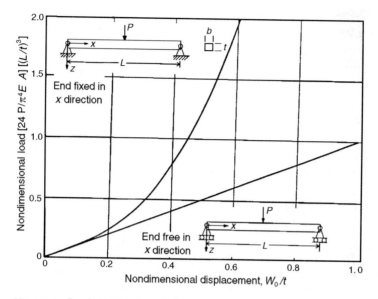

Figure 3.10 Load–displacement behavior of a simply supported beam under a concentrated central load.

that plane sections remain planar, then the strain in the x direction at any point in the beam can be written as

$$\varepsilon_x = \frac{du}{dx} + \frac{1}{2}\left(\frac{du}{dx}\right)^2 + \frac{1}{2}\left(\frac{dw}{dx}\right)^2 + zK \qquad (3.17)$$

where K is the curvature of the neutral axis of the deformed beam and is defined in terms of the lateral displacement w as

$$K = \frac{-\dfrac{d^2w}{dx^2}}{\left[1+\left(\dfrac{dw}{dx}\right)^2\right]^{3/2}} \qquad (3.18)$$

The first derivative of the lateral displacement (dw/dx) represents the local rotation of the beam. Because $(dw/dx)^2$ is generally much less than 1, the curvature can usually be accurately approximated as

$$K \approx -\frac{d^2w}{dx^2} \qquad (3.19)$$

Furthermore, because du/dx is much less than dw/dx in laterally loaded beams, Eq. (3.17) can be simplified to

$$\varepsilon_x = \frac{du}{dx} + \frac{1}{2}\left(\frac{dw}{dx}\right)^2 + zK \tag{3.20}$$

Equation (3.20) includes the square of the beam rotation, $(dw/dx)^2$, again. However, although neglecting it in Eq. (3.18) only required that it be small in comparison to 1, the linearization of Eq. (3.20) requires that $(dw/dx)^2$ be much less than du/dx. Since du/dx is much less than 1, this is a much more stringent requirement. Because there are two quantitatively different assumptions with regard to the size of rotations, the term "moderately large" is sometimes used to describe the situation where $(dw/dx)^2$ is approximately the same size as du/dx. If the rotations are small [$(dw/dx)^2$ much less than du/dx)], then Eq. (3.20) can be linearized to

$$\varepsilon_x = \frac{du}{dx} + zK \tag{3.21}$$

which is the well-known strain–displacement relation for linear beam theory and allows uncoupled solutions for the u and w displacements.

Now consider the boundary conditions at the ends of the beam. If one or both ends of the beam are assumed to be free to move in the x direction, then there can be no net force F in that direction. Using linear elasticity to relate stress and strain and then integrating the x component of stress over the cross section of the beam, the resulting force is

$$F = E\int_A\left[\frac{du}{dx} + \frac{1}{2}\left(\frac{dw}{dx}\right)^2 - z\frac{d^2w}{dx^2}\right]dy\,dz = EA\left[\frac{du}{dx} + \frac{1}{2}\left(\frac{dw}{dx}\right)^2\right] = 0 \tag{3.22}$$

where A is the cross-sectional area of the beam and the rotations are allowed to become moderately large. Using Eq. (3.22) along with Eq. (3.20), the governing strain–displacement relation for the beam under a central, lateral load and the above-mentioned boundary conditions is

$$\varepsilon_x = +zK \tag{3.23}$$

Therefore, under the boundary condition of free displacement in the x direction at one or both ends of the beam, the full, nonlinear strain–displacement relation given in Eq. (3.20) reduces to the linear equation given in Eq. (3.23) without the requirement that $(dw/dx)^2$ be much less than du/dx. Using Eq. (3.23) and then applying the equations of equilibrium results in the standard linear equation for beam theory.

The situation is different, however, if both ends of the beam in Fig. 3.10 are prevented from moving in the x direction. Under this set of boundary conditions, it is no longer possible to equate the force in the x direction of the beam to zero. As a result, Eq. (3.22) is no longer available and Eq. (3.20) retains its full nonlinear form. In this case, unless the requirement that $(dw/dx)^2$ be much less than du/dx is applicable, the governing differential equations for equilibrium will be nonlinear. From a physical point of view, the bending and stretching deformation of a beam become increasingly coupled if the ends of the beam are restrained in the x direction and the rotations become moderately large; that is, $(dw/dx)^2$ is approximately the same size as du/dx.

Although an exact solution to these equations is not straightforward, an approximate solution can be obtained by using the theory of minimum total potential energy and approximations for the displacements that fulfill the displacement boundary conditions. Using simple trigonometric expressions, the first approximations for the lateral and in-plane displacements w and u are

$$w = W_0 \sin \frac{\pi x}{L} \tag{3.24}$$

$$u = U_0 \sin \frac{2\pi x}{L} \tag{3.25}$$

where W_0 and U_0 are unknown coefficients that must be determined through minimizing the total potential energy of the system with respect to these unknown displacements. The total potential energy of the system can be expressed as

$$U + V = \int_0^L \int_{-t/2}^{t/2} \frac{Eb}{2} \varepsilon_x^2 \, dz \, dx - PW_0 \tag{3.26}$$

Differentiating the total potential energy expression (3.26) with respect to W_0 and U_0 and setting both expressions equal to zero results in two nonlinear, algebraic equations for W_0 and U_0. Using these two equations, the load–displacement behavior of the fully nonlinear problem can be derived as

$$P = \frac{\pi^4 EI}{2\,L^3} W_0 + \frac{\pi^4 EA}{4\,L^3} W_0^3 \tag{3.27}$$

In Eq. (3.27), the linear term in W_0 can be recognized as the approximate relationship between load and central beam displacement arising from the standard Euler–Bernoulli equation for linear beam theory. The cubic term provides additional stiffening to the system as the displacements become large. This stiffening is physically due to stretching of the beam's length dimension as a result of lateral displacements perpendicular to the original length of the beam. As can be seen in Fig. 3.10, this nonlinear effect upon the lateral stiffness of the beam becomes significant after the displacement of the beam becomes large with respect to the beam thickness ($W_0/t \sim 0.3$). This in turn leads to the terminology of describing these results as a "large-displacement" solution; that is, displacements significant with respect to the thickness of the beam can be accurately accommodated.

For most realistic beam structures, the boundary conditions at the ends are more likely to be approximated by the set of boundary conditions allowing free displacement in the x direction. Effective fixity of the u displacements at the ends is difficult to accomplish practically. As a result, such problems can usually be treated very accurately with normal, linearized beam theory, even when the square of the rotations, $(dw/dx)^2$, becomes large with respect to $|du/dx|$. Williams gives an example of the effect of very large rotations, $(dw/dx)^2 \sim 1.0$, on prediction of lateral deflection for a beam with unrestrained ends in Ref. 13. Significant differences between this "very large" rotation solution and predictions based upon small rotation theory only appear when lateral displacements approach 10% to 20% of the beam *length*. These are much larger displacements than those associated with "moderate rotation" theory. As a result, using the fully linearized Eqs. (3.18) and (3.20) to predict the deformation and stress in plastic beams should generally be accurate for engineering problems even though the low modulus of these plastic materials will produce large displacements in comparison to the *thickness* of a beam.

In the previous example we saw that the nonlinear rotational terms in the strain–displacement relations can lead to apparent stiffening in the load–displacement behavior of transversely loaded beams. In the situation examined, the beam's in-plane boundary conditions led to tensile in-plane strains and stresses that added to the customary linear bending stiffness. It is important to understand that in some circumstances, these same nonlinear rotational terms can also lead to a reduction of stiffness as displacement increases. In many cases the important difference in the character of the problem is that the in-plane stresses and strains are compressive in nature when the softening phenomenon occurs.

Consider, for example, the simple beam column shown in Fig. 3.11. For illustration purposes, assume that the beam is simply supported at both ends. The lower support in this example is fixed in both coordinate directions. The upper support is permitted to displace in the x-coordinate direction, but is prohibited from moving in the z-coordinate direction. At the upper support, a load P is applied to a short rigid link of length e. The force is in the negative x direction. As in the previous examples of large nonlinear rotations of beams and plates, the load–displacement behavior of this structure can be approximated using the theory of minimum total potential energy. The total potential energy for this problem can be written as:

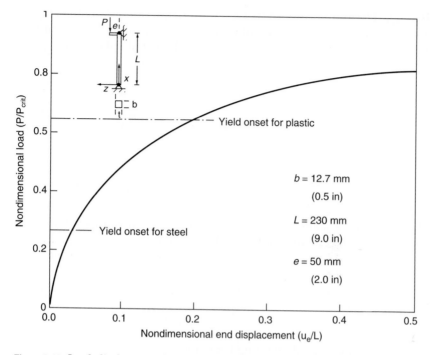

Figure 3.11 Load–displacement behavior of an eccentrically loaded beam column.

$$U + V = \int_0^L \frac{EA}{2}(\varepsilon_0)^2 dx + \int_0^L \frac{EI}{2}(K)^2 + Pu_e \qquad (3.28)$$

where

$$\varepsilon_0 = \frac{du}{dx} + \frac{1}{2}\left(\frac{dw}{dx}\right)^2 \qquad (3.29a)$$

$$K = -\frac{d^2w}{dx^2} \qquad (3.29b)$$

$$u_e = u(L) - e\frac{dw}{dx}(L) \qquad (3.29c)$$

The u and w displacements can be approximated as:

$$u(x) = \frac{U_0 x}{L} + U_2 \sin\left(\frac{2\pi x}{L}\right) \qquad (3.30a)$$

$$w(x) = W_1 \sin\left(\frac{\pi x}{L}\right) \qquad (3.30b)$$

where U_0, U_2, and W_1 are unknown. Minimizing the total potential energy with respect to the three unknown displacements, U_0, U_2, and W_1, leads to three nonlinear equations in the unknown displacements:

$$U_0 = -\frac{\pi^2}{4}\frac{W_1^2}{L} - \frac{PL}{EA} \tag{3.31a}$$

$$U_2 = -\frac{\pi}{8}\left(\frac{W_1^2}{L}\right) \tag{3.31b}$$

$$\frac{1}{2}\frac{\pi^2 EA}{L^2}U_0 W_1 + \frac{\pi^3 EA}{2L^2}U_2 W_1 + \frac{3}{16}\frac{\pi^4 EA}{L^3}W_1^3 + \frac{\pi^4 EIW_1}{2L^3} + \frac{P\pi e}{L} = 0 \tag{3.31c}$$

Substituting for U_0 and U_2 in Eq. (3.31c) leads to an expression for the transverse displacement magnitude W_1 in the form

$$W_1 = -\frac{\left(\dfrac{2e}{\pi}\right)\left(\dfrac{P}{P_{\text{crit}}}\right)}{1 - \left(\dfrac{P}{P_{\text{crit}}}\right)} \tag{3.32}$$

where P_{crit} is the critical buckling load of a column expressed as

$$P_{\text{crit}} = \frac{\pi^2 EI}{L^2} \tag{3.33}$$

Using Eq. (3.29c), the expression for the displacement under the load, P, can be written as

$$u_e = -2L\left(\frac{P}{P_{\text{crit}}}\right)\left(\frac{e}{L}\right)^2\left[\frac{1}{1 - \dfrac{P}{P_{\text{crit}}}} + \frac{\pi^3}{4}\left(\frac{I}{Ae^2}\right)\right] - \frac{1}{\pi^2}\left(\frac{e}{L}\right)^2\left[\frac{\dfrac{P}{P_{\text{crit}}}}{1 - \left(\dfrac{P}{P_{\text{crit}}}\right)}\right]^2 \tag{3.34}$$

The negative sign in Eq. (3.34) simply indicates that the direction of displacement in Fig. 3.11 is in the negative x direction. Equation (3.34) can be used to create the load-versus-displacement curve shown in Fig. 3.11. As can be seen from the figure, as the load increases, the stiffness of this structure steadily decreases and displacements become unbounded as the load approaches the column's critical buckling load. It should be emphasized that there is no material nonlinearity included in analyzing this problem. The only nonlinearity is the geometric effect of large rotations in the beam.

As an illustration of why this nonlinear behavior can be more significant for plastics than other materials, we can consider two structures like the one shown in Fig. 3.11: one made of steel, and one made of polycarbonate. In order to compare their performance on an equal basis, the thickness of the plastic beam will be chosen in order to ensure that the plastic structure has the same stiffness as the steel structure. For a steel modulus of 210 GPa (30×10^6 psi) and a beam thickness of 1.1 mm (0.043 in), the thickness of a beam made of polycarbonate with a modulus of 2.1 GPa (30×10^4 psi), must be 5.1 mm (0.2 in) in order to supply equal stiffness. The other geometric parameters governing this problem are assumed to be the same for both steel and plastic beams and are listed in Fig. 3.11. Using these parametric values, the nondimensional load–displacement relationship for this structure is shown in the figure. The curve is equally valid for both steel and polycarbonate. Also noted in the figure are the nondimensional loads and displacements at which structures made of steel and plastic would first reach a yield condition. As can be seen, for this geometry, the steel structure yields in the nearly linear range of the problem, well below the critical buckling load of the beam. Linear strain–displacement assumptions would have been accurate for this structure. The plastic structure, on the other hand, experiences considerable nonlinearity due to the large rotations in the strain–displacement equations before it reaches yield anywhere in the beam.

In addition to illustrating the importance of geometric nonlinearities to accurate modeling of plastic structural behavior, Fig. 3.11 also illustrates another characteristic of certain plastic structures. Since the energy absorbed in this structure is simply the area under the load–displacement curve in Fig. 3.11, the plastic structure would sustain much higher loads and absorb much more energy before yielding than the steel structure of comparable stiffness.

Although the assumptions associated with thin-plate theory are the same as for beam theory, the conditions under which these assumptions are effective are very different because of the two-dimensional geometry of flat-plate structures. The expressions for strains in terms of displacements for flat plates are recognizable extensions of the beam expression given in Eq. (3.20). There are now two direct strains that must be treated as well as a shear strain and they can be written as

$$\varepsilon_x = \frac{\partial u}{\partial x} + \frac{1}{2}\left(\frac{\partial w}{\partial x}\right)^2 + zK_{xx} \tag{3.35a}$$

$$\varepsilon_y = \frac{\partial v}{\partial y} + \frac{1}{2}\left(\frac{\partial w}{\partial y}\right)^2 + zK_{yy} \tag{3.35b}$$

$$\gamma_{xy} = \frac{\partial u}{\partial y} + \frac{\partial v}{\partial x} + \left(\frac{\partial w}{\partial x}\right)\left(\frac{\partial w}{\partial y}\right) + 2zK_{xy} \tag{3.35c}$$

Here, K_{xx}, K_{yy}, and K_{xy} are curvatures that can be approximated as

$$K_{xx} \approx -\frac{\partial^2 w}{\partial x^2} \tag{3.36a}$$

$$K_{yy} \approx -\frac{\partial^2 w}{\partial y^2} \tag{3.36b}$$

$$K_{xy} \approx -\frac{\partial^2 w}{\partial x\, \partial y} \tag{3.36c}$$

As for beams, the small rotation assumption can be obtained by neglecting the quadratic rotation terms in Eqs. (3.35a), (3.35b), and (3.35c).

A significant difference in the range of problems where small rotation theory is effective appears for plates in comparison to the beams previously discussed. Because of the two-dimensional nature of flat plates, the simple equilibrium equation (3.22) used for beams that have ends unrestrained in the in-plane direction is no longer available. As a result, the full, nonlinear equations of equilibrium in terms of the lateral displacement w and the two in-plane displacements u and v remain coupled even if the edges of the flat plate are completely unrestrained in the in-plane directions. Therefore, while small-rotation beam theory is very accurate for moderately large rotations, as long as the beam ends are unrestrained in the neutral-axis direction, the same cannot be said for flat plates.

Realistically sized flat plates supported against lateral displacements on their entire periphery show significant nonlinear behavior when displacements approach the value of the plate thickness even if the plate edges are completely unrestrained in the in-plane directions. If there is in-plane restraint applied at the edges, then the range of displacements over which small-rotation theory is effective becomes even more restricted. Figure 3.12 illustrates the nonlinear response of a transversely loaded plate and its dependence upon the boundary conditions associated with the in-plane displacements of the plate. The plate considered in Fig. 3.12 is simply supported (i.e., free to rotate but not to displace laterally) at its edges and subjected to a uniform lateral pressure. Four load–deflection curves are plotted in the figure. The same classical, simply supported conditions of zero lateral displacement and zero moment are applicable to all four curves. However, each curve is associated with a different set of boundary conditions for the in-plane displacement and force variables. The lowest of the three curves represents the classical, small-displacement solution to the problem. The curve just above the classical results reflects a nonlinear

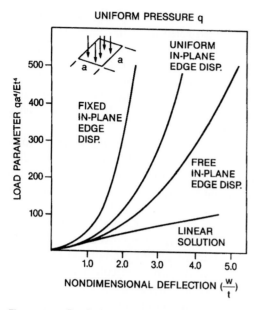

UNIFORM PRESSURE q

Figure 3.12 Load–displacement behavior of a flat plate loaded under uniform pressure for various in-plane boundary conditions.

solution to the problem where the edges of the plate are completely free of in-plane restraint. The in-plane boundary conditions for the next curve allow uniform edge displacement perpendicular to each edge, but prohibit any relative displacement along an edge. Physically, this is equivalent to "framing" the plate with a stiffener that is very rigid with respect to the in-plane variables u and v. The uppermost of the four curves is associated with complete displacement restraint at the edges. All of the three displacements u, v, and w, are forced to be zero along the edges in this case. All of the situations examined in Fig. 3.12, including the plate with free in-plane displacements, show noticeable nonlinear stiffening as the lateral displacement exceeds the value of the plate thickness.

Physically, the stiffening effect evident in Fig. 3.12, even for free edges, is a result of the fact that a flat surface cannot be deformed into a doubly curved surface without increasing its surface area—that is, without stretching the surface. In contrast, a flat surface can be deformed into a cylinder (singly curved surface) with only bending deformations. A simple experiment illustrating this point can be carried out with a piece of paper. Although the paper can be easily rolled up into a cylinder, it cannot be wrapped around a sphere without wrinkling or tearing the paper. Therefore, if all four edges of a plate are restrained in the lateral (w) direction, then small-rotation, thin-plate theory will only be accurate as long as this surface stretching effect is negligible. If, on the other hand, only one edge or two opposite edges are restrained laterally, then the physical situation is more similar to that of a beam, because the plate can take on a cylindrical deformation without stretching its midsurface. In such a case, small rotation theory again has a wider range of applicability.

As in the previous example of the eccentrically loaded beam column, it is of interest to address the question of why the nonlinear, moderately large rotation range of this theory should be any more important for plastic materials than it would be for metals. To address this issue, let us consider the simply supported plate subjected to a uniform lateral pressure again. For this example we will consider the edges of the plate to be completely unrestrained in the in-plane directions. After solution of this problem (the finite-element technique is used here), the maximum strain at the center of the plate can be calculated and is plotted in Fig. 3.13 as a function of the center-plate deflection, nondimensionalized by the plate thickness. At this stage of the discussion, no assumption about the plate material has entered. The nondimensionalized behavior shown in Fig. 3.13 is independent of material properties. First, let Fig. 3.13 be interpreted under the assumption that the plate

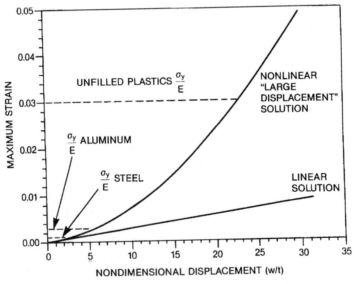

Figure 3.13 Tensile strain at the bottom surface of a uniformly loaded rectangular plate as a function of nondimensional maximum displacement.

is made of steel with a yield stress of 280 MPa (41×10^3 psi) and a Young's modulus of 210 GPa (30×10^6 psi). If, in addition, we impose the practical engineering constraint requiring strains to be less than the material's yield strain, then for steel it can be seen that the linear portion of the curve in Fig. 3.13 associated with small-rotation theory is entirely adequate. If we interpret Fig. 3.13 in terms of aluminum, with a yield stress of 210 MPa (30×10^3 psi) and Young's modulus of 70 GPa (10×10^6 psi), then nonlinear effects of large rotations do become evident before yield, but the linear theory still has a wide range of applicability. In contrast, however, if Fig. 3.13 is used to assess the behavior of a plastic material with a yield stress of 70 MPa (10×10^3 psi) and a modulus of 2.1 GPa (30×10^4 psi), then it is strikingly clear that there is an extremely broad range of structural behavior where the material would be completely linear-elastic but where linear, small-rotation plate theory is completely inadequate. This comparison makes it clear that nonlinear, moderately large-rotation solutions to thin-plate problems can be extremely important for efficient design with plastics. If linear small-rotation theory is used, then it is very possible to calculate displacements that are significantly larger than those that will actually occur under a given lateral load.

Unfortunately, analytically predicting plate deformation using nonlinear, moderately large-rotation plate theory is not as straightforward a task as using the linear theory. Many calculations based on the moderately large-rotation theory have been published. Reference 14 summarizes some of the early examples, but they are not in a form that makes them as easy to use as the extensively tabulated results of small-rotation theory. In some cases it may be efficient to use approximate solutions to these nonlinear problems generated for a few important geometries and boundary conditions by such techniques as minimization of the total potential energy, as applied earlier. These solutions are compatible with use on personal computers and provide for quick initial design studies. In situations where the geometry is more complex and/or a higher degree of accuracy is required, it may be necessary to use a finite-element program with the capability to handle moderately large-rotation plate problems. Programs capable of this task are widely available and have become much easier to use. If lateral loading of flat plates is a problem routinely encountered in a particular engineering business, it is possible to create additional computer software treating specific geometries and making the preparation of data and the interpretation of results from finite-element programs extremely efficient, as has been reported in Ref. 15.

One realistic and geometrically simple example where large deflections of plastic plates are encountered is in glazing applications such as those shown in Fig. 3.14. Large sheets of plastic such as these can ex-

Figure 3.14 Polycarbonate sheet used in window applications.

perience distributed loads from wind or snow. In Ref. 15, the application of nonlinear, large-displacement analysis is discussed relative to the accuracy of predicting the deflections of large polycarbonate sheets using the finite-element approach and accounting for the effects of large displacements. Figure 3.15 reports results from that reference for two different sheet geometries of polycarbonate. Both sheets were 6.35 mm (0.25 in) thick. One sheet measured 1.22 m × 2.44 m (48 in × 96 in) while the other measured 1.22 m × 1.22 m (48 in × 48 in). In both cases, the sheets were supported between metal plates at the plate perimeter and then loaded by uniform pressure over the entire unsupported area of the plate. Figure 3.15 illustrates the analytically predicted pressure-versus-deflection curve for the two geometries, as well as experimental data from tests on these two geometries also. As can be seen, the response of the sheet under pressure is extremely nonlinear in nature. However, using the large-rotation (large-displacement) analysis, the numerical results very accurately predict the deflections of the polycarbonate sheet. As an additional indication of the importance of the nonlinear geometric effects in this analysis, the linear prediction for the 1.22 m × 2.44 m (48 in × 96 in) sheet is also presented. As can be seen, the prediction is extremely inaccurate in this application.

Nonlinearity due to the load–deformation interaction

A third type of nonlinear behavior is the result of the interaction of deformation with application of load. Linear analysis assumes that the

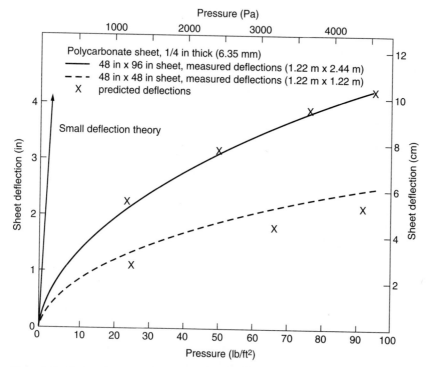

Figure 3.15 Comparison of experimentally measured and analytically predicted load–displacement behavior of polycarbonate glazing under uniform pressure.

location and distribution of load in a problem do not change during deformation. This assumption may not always be valid, especially when deformations become large. A good example of such a problem is the indentation of a plastic disk by a hemispherical steel indenter, illustrated schematically in Fig. 3.16. This geometry is encountered in an impact test commonly used to rate thermoplastic materials and it will be discussed more thoroughly in subsequent chapters. Here we will concentrate on the nonlinear application of load in this problem.

When the indenter first contacts the plastic disk, the load can be assumed to be applied in a concentrated fashion at the disk's center. Linear solutions to such a load application exist and the stiffness is defined in Refs. 14 and 16. However, as the indenter displacement increases, the load is distributed over an increasing area of the disk. The exact distribution of that load cannot be defined by the analyst. It is literally part of the solution.

Many modern FEA systems (Refs. 9 to 11, for example) offer methods of addressing such problems of varying contact between two bodies. In

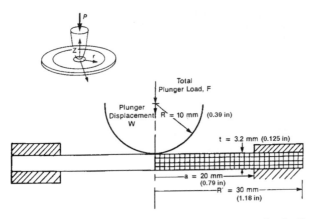

Figure 3.16 Schematic illustration of an axisymmetric plastic disk loaded by a hemispherical steel indenter.

most cases, some general zone of expected contact must be defined by the analyst. This definition simply defines where the software should search for potential contact. The bodies undergoing contact may be defined as rigid or flexible in nature.

In the case of the indenter and disk, the steel indenter can be assumed to be rigid in comparison to the plastic disk. The potential areas of contact in this problem include the indenter surface and the central part of the disk. In this case, an area slightly larger than the projected area of the indenter on the disk is adequate for definition. The finite elements employed in this analysis are 2-D axisymmetric elements.

Using this definition, Fig. 3.17 illustrates the difference in two predictions—one based upon a linear-elastic analysis of a polycarbonate disk subjected to a concentrated central load and the second based upon a more general, nonlinear analysis that addresses the changing distribution of load during the disk's deformation. It should be emphasized that the contact nature of the load is the only nonlinearity treated in this problem. In both cases, the material is assumed to behave in a linear-elastic fashion with the properties of polycarbonate, and small-rotation continuum theory is applied. As can be seen in Fig. 3.17, the initial stiffness predicted by both methods is the same. However, as the indenter displacement increases, the nonlinear analysis, which accounts for the changing contact area, predicts a stiffer response than the linear analysis. It should be emphasized that the increased stiffness in this problem, due to accounting for load changes properly as a function of deformation, should not be viewed as a general characteristic of this nonlinearity. In other geometries, it is equally possible that this type of nonlinearity could result in reduced stiffness as deformation increases.

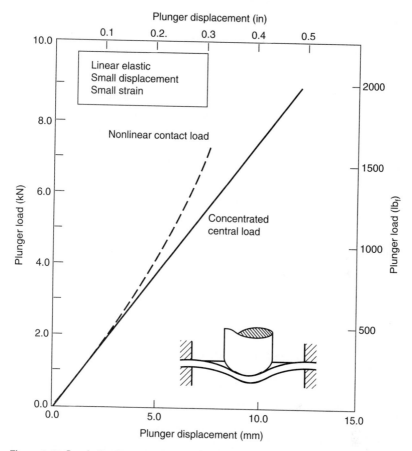

Figure 3.17 Load–displacement behavior as predicted by linear theory and contact–load simulation (linear material and small-rotation assumptions).

Closure

The finite-element technique that has been outlined briefly in this chapter is a very useful analysis tool for parts made of thermoplastic. It provides the capability to analyze complex part geometries as well as nonlinear behavior.

There are several different types of nonlinearity that may be encountered during analysis of thermoplastic parts. The constitutive equations relating stress to strain may be nonlinear in nature, as they are for ductile metals. A second type of nonlinearity can be encountered in the equations relating strain to displacement. This nonlinearity is especially significant in beam, plate, and shell structures, when rotations associated with transverse displacements become moderately large.

The behavior associated with this nonlinearity can be either stiffening or softening in its nature. Because of the large strains that plastics are capable of sustaining before incurring damage, this latter type of nonlinearity is particularly important. Finally, many modern nonlinear finite-element codes are also capable of accounting for the nonlinear nature of loads associated with the contact of two bodies.

Throughout the rest of this text, the finite-element analysis technique will be applied to analyze the behavior of numerous thermoplastic component and test specimen problems. In many cases it would be impossible to account for some of the very important behavior without this approach. Although the discussion of this approach in this chapter has been simplistic, it will provide an initial basis for understanding the results that follow.

References

1. O. C. Zienkiewicz, *The Finite Element Method*, McGraw-Hill, London, 1977.
2. K. J. Bathe and E. Wilson, *Numerical Methods in Finite Element Analysis*, Prentice-Hall, Englewood Cliffs, NJ, 1976.
3. K. J. Bathe, *Finite Element Procedures in Engineering Analysis*, Prentice-Hall, Englewood Cliffs, NJ, 1982.
4. G. Carey and J. T. Oden, *Finite Elements: An Introduction*, Prentice-Hall, Englewood Cliffs, NJ, 1983.
5. T. J. R. Hughes, *The Finite Element Method*, Prentice-Hall International, Englewood Cliffs, NJ, 1987.
6. T. J. R. Hughes and E. Hinton, *Finite Element Methods for Plate and Shell Structures*, Vol. 1, Pineridge Press International, Swansea, U.K., 1986.
7. John Robinson, Early FEM Pioneers, Robinson and Associates, Dorset, England, 1985.
8. *ANSYS® Engineering Analysis System Theoretical Manual*, P. C. Kohnks, Ed., Swanson Analysis Systems, Houston, PA, 1989.
9. *ADINA® Theory and Modelling Guide*, Report No. ARD-87-4, ADINA R&D, Inc., Watertown, MA, 1987.
10. *ABAQUS® User's Manual*, Ver. 4.7, Hibbitt, Karlsson and Sorensen, Providence, RI, 1988.
11. *MARC® General Purpose Finite Element Program*, Vol. A, *User Information Manual*, MARC Analysis Research, Palo Alto, CA, 1983.
12. G. G. Trantina and M. D. Minnichelli, "The Effect of Nonlinear Material Behavior on Snap-Fit Design," *Proceedings of the 1987 SPE Annual Technical Meeting*, Society of Plastics Engineers, Brookfield Center, CT, 1987, pp. 438–441.
13. J. G. Williams, *Stress Analysis of Polymers*, Wiley, New York, 1980, p. 160.
14. S. Timoshenko and S. Woinowsky-Krieger, *Theory of Plates and Shells*, McGraw-Hill, New York, 1959, pp. 396–425.
15. M. D. Minnichelli, C. M. Mulcahy, and G. G. Trantina, "Automated Structural Analysis of Plastic Sheet," *Advanced Polymer Technology*, 6 (1): 73–78, 1986.
16. R. J. Roark, *Formulas for Stress and Strain*, McGraw-Hill, New York, 1965, p. 217.

4

Stiffness

Structural stiffness is one of the most fundamental engineering design criteria for components made of any material. Polymeric materials are no exception to this rule. In fact, since the Young's modulus of most polymers is low in comparison to many other structural materials, stiffness is often a very significant concern with respect to effective use of a plastic material.

As discussed in the Introduction, loads on a structural component can be categorized into several significant and reasonably distinct groups. For the discussion relative to stiffness in this chapter, the loads will generally be considered to be applied slowly enough so that the inertial properties of the structural component are not a significant issue in the structural response. Furthermore, the discussion in this chapter will also be limited to situations where the load is removed over a period of time comparable to its application time, as shown in Fig. 4.1. There are circumstances in which loads are applied and maintained on a structure over substantial periods of time. In such situations, the time-dependent response of the material can be significant. This issue will be addressed in a separate chapter.

It should be emphasized from the outset that the stiffness of a structural component is dependent upon both material properties and part geometry. Although the material stiffness of polymers is comparably low, adequate structural stiffness for a component can often be attained through appropriate use of geometry. For example, in spite of the fact that most engineering thermoplastics have moduli two orders of magnitude less than steel, a thermoplastic box beam can provide adequate stiffness to replace typical open section steel bumper beams. In so doing it is possible to take advantage of other desirable characteristics of plastics, such as corrosion resistance or improved manufac-

Load

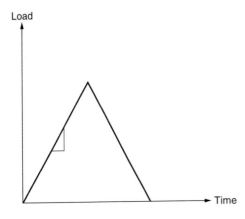

Figure 4.1 Static loading followed by unloading

Time

turability and low cost. Often the range of geometry available to the designer is also closely related to the manufacturing process used to create the part. A particular geometry available when a part is injection molded may not be a realistic possibility if a blow-molding process is used instead. During the course of this chapter, issues related to both material and geometric contributions to structural stiffness will be discussed.

The contribution of the material to a part's stiffness is usually defined by measuring the relationship between stress and strain. Practically speaking, this is most often accomplished by testing simple tensile specimens while measuring applied load and induced strain. The size and shape of these specimens have been standardized by the American Society for Testing of Materials (ASTM),[1] as illustrated for two acceptable geometries in Fig. 4.2. In a typical test such as the one shown in Fig. 4.3, a tensile load is applied to the specimen and measured with a load cell. Because of the regular geometry of the central portion of the specimen, the stress and strain in that region are assumed to be uniform. The stress is calculated using the simple formula

$$\sigma = P/A \qquad (4.1)$$

where A is the cross-sectional area of the test specimen. The strain induced in the specimen is usually measured with an extensometer mounted in the central portion of the specimen. If the strain field in the specimen is nonuniform for any reason, such as if strain localization takes place—a subject that will be discussed in the next chapter—then both the stress and the strain measurement techniques must be altered for proper definition of the material behavior. For the design analyst, these measurements relating stress to strain are absolutely essential. They are the most fundamental portion of the bridge be-

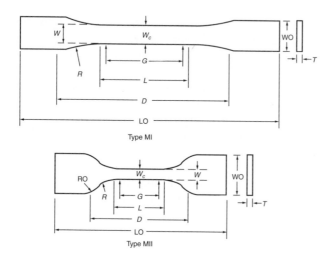

Type MI

Type MII

	Dimensions in mm	
	Type MI	Type MII
W—Width of narrow section*	10	6
L—Length of narrow section	60	33
WO—Width overall	20	25
LO—Length overall	150	115
G—Gage length	50	25
D—Distance between grips	115	80
R—Radius of fillet	60	14
RO—Outer radius	. . .	25
T	4	4

* The width at center W_c shall be plus 0.0 mm and minus 0.1 mm compared with W at other parts of reduced section.

Figure 4.2 ASTM tensile specimen.

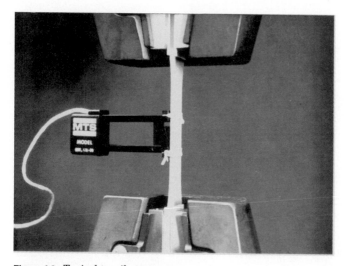

Figure 4.3 Typical tensile test setup.

tween the world of materials and an engineered part. Analysts must be confident that these values quantifiably characterize the materials in question as they define the compatible geometry required to ensure component performance.

The general form of the complete stress–strain relationship for plastic materials may vary significantly. Two contrasting examples are shown in Fig. 4.4. Because of the very large deformations encountered in polycarbonate, a deformation variable commonly used to define large deformations is used in the abscissa of this figure. This parameter is sometimes referred to as "stretch" and is defined as $\lambda = l/l_0$ where l is the current length of a material segment whose original length was l_0. For small values of stretch this parameter is related to engineering strain ε as $\varepsilon = \lambda - 1$. Unfilled polycarbonate has been observed to accommodate stretches approaching 2.0 before failure and over that range displays a very nonlinear relationship with stress. Other plastics, like the glass-filled polycarbonate, also shown in Fig. 4.4, incur values of stretch only slightly larger than 1 ($\lambda \approx 1.03$) before failure and exhibit a much more linear relationship between stress and stretch. From a material response standpoint, this chapter will focus upon the initial, approximately linear portion of the stress–stretch re-

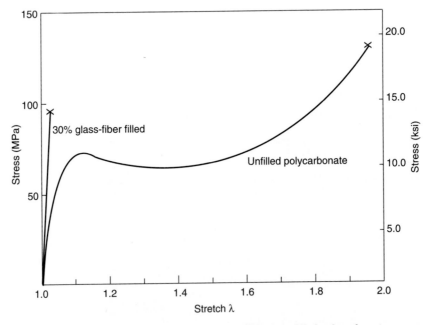

Figure 4.4 Stress–stretch curves for unfilled and 30% glass-filled polycarbonate.

lationship for plastic materials. In this regime, strain ($\varepsilon = \lambda - 1$) is more commonly used to describe deformation. Some of the consequences of the very nonlinear behavior beyond 5% strain exemplified by a plastic such as polycarbonate, as shown in Fig. 4.4, will be discussed in a subsequent chapter.

Even within the initial region of the stress–strain curve, there are several significant differences between the general properties of plastics and those of other engineering materials. Young's modulus for plastics, determined from the initial slope of the stress–strain curve, is usually one to two orders of magnitude less than that of most engineering metals. In comparison to the 207 GPa (30×10^6 psi) modulus of steel, for example, polycarbonate offers a modulus of 2.07 GPa (30×10^4 psi). This fact rather naturally leads to the appearance of large deformations in many plastic parts unless structural shape is inventively used to provide overall component stiffness.

Fortunately, one of the greatest advantages of plastics is their ability to be easily and economically formed into complex structural shapes. Although plastic parts must usually take the form of thin-walled structures, a beam made of thermoplastic can be made very stiff simply by molding it into the form of a hollow box-section beam, like that shown in Fig. 4.5. Similarly, a flat plastic panel may not be stiff enough in a thin-sheet form, but it can be easily molded to include extensive stiffening ribs, like the panel shown in Fig. 4.6. Alternatively, the blow-molding process may be used to enhance stiffness in two-dimensional structures, such as the one shown in Fig. 4.7—a double-walled panel that has been filled with foam.

Two additional differences between plastics and other common engineering materials that are noticeable in Fig. 4.4 are the large-strain limit for the initial, approximately linear portion of the stress–strain curve, and the generally small-stress limit. The linear-elastic strain limit for a steel is on the order of 0.001 to 0.002. A typical unfilled thermoplastic will experience over an order of magnitude more strain, on the order of 0.02, before undergoing major nonlinearity. As discussed in Chap. 3, this behavior leads to a number of nonlinear effects that are geometrically based. Several additional examples of this behavior will be discussed subsequently in this chapter.

A final difference between plastics and other engineering materials worth highlighting at this stage is one that cannot be overtly observed in the stress–strain curves of the materials, although its effect is certainly related to the concept of stress–strain behavior. The morphology or microstructural form of plastics is obviously very different from metals and other engineering materials. In some cases, this material difference does not play a significant role in understanding the stiffness of a plastic component. However, there are plastic materials whose micro-

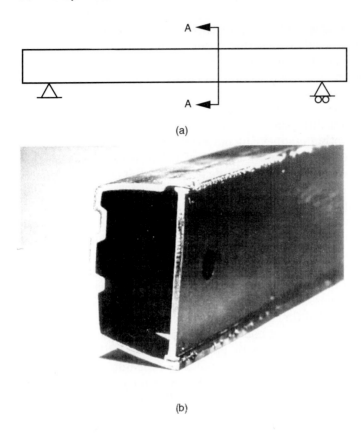

(a)

(b)

Figure 4.5 (a) Simply supported beam; (b) box-beam cross section (A–A).

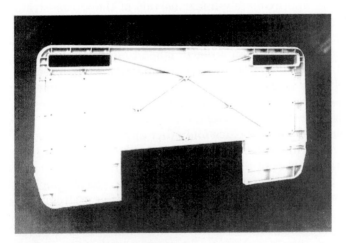

Figure 4.6 Panel stiffened with ribs.

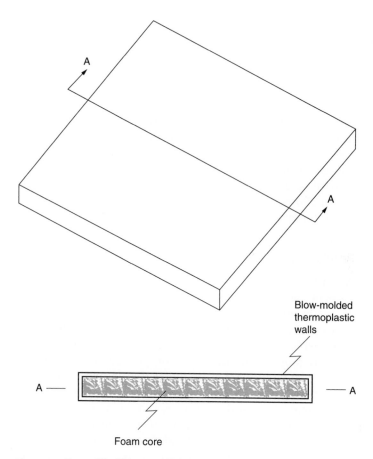

Figure 4.7 Foam-filled blow-molded panel.

structure is nonhomogeneous on a scale much larger than that which engineers are generally familiar based upon their experience with metals. Structural foam and compression-molded thermoplastic-impregnated glass mat sheets are good examples of this behavior. This nonhomogeneity can lead to fundamental issues related to the measurement of stress–strain information that is useful in an engineering sense.

There will be two distinct subjects dealt with in this chapter on stiffness. In the first subsection, several fundamental issues related to the application of engineering stress–strain data describing the stiffness of plastic materials will be discussed. In some cases it will be shown that standard approaches to measuring stress–strain relationships can be used to generate data that are quite useful in predicting component stiffness. Several contrasting examples will also be presented in which the characteristic morphology of some plastic materials leads to fundamental inaccuracy and confusion if standard approaches are followed.

The second subsection in this chapter will deal with the general subject of using geometry and the processing opportunities available with plastics to achieve enhanced structural stiffness. Several different concepts for attaining structural rigidity will be examined, as well as the engineering predictive models that can be used to consider their efficiency in providing stiffness. In both subsections, specific practical examples of thermoplastic parts will be provided to illustrate both the characteristic behavior of interest and the ability to model and predict that behavior in an engineering sense.

Issues of Material Stiffness

The ability to quantify a material's intrinsic material stiffness is essential to the process of design analysis. It should be emphasized that this measured stiffness must be a material property in the sense that its value is independent of geometry. Only material properties can be used to predict how an arbitrary structure will perform. Although stiffness is usually one of the more straightforward properties of a material to measure, there are issues that must be addressed in order to build a consistent process that can be applied in engineering design.

There are a number of general issues, including homogeneity and isotropy, that should be considered relative to this process of establishing material properties that quantify stiffness. A material is described as homogeneous if its microstructure or morphology does not vary with space. Most materials can be described as nonhomogeneous on some scale, but usually this scale is so small as to be inconsequential from an engineering standpoint. However, for material nonhomogeneity on a scale equivalent to design dimensions, nonstandard approaches to measuring stiffness may be required.

The term *anisotropic* is used to describe a material whose stiffness properties are dependent on direction at any given point in a material. Wood is a good example of an anisotropic material, since its stiffness and strength properties are very different in the direction of the wood grain in comparison to the direction perpendicular to the grain. In contrast, the material properties of an isotropic material are independent of direction. For a homogeneous and isotropic material, two properties are sufficient to define material stiffness. Usually Young's modulus and the Poisson ratio are the properties measured. There are a variety of subclassifications of anisotropy and a complete discussion of this subject can be found in textbooks on elasticity such as Ref. 2. It can be said in general, however, that if a material is anisotropic, more than two material properties must be identified to define material stiffness adequately.

In the discussion that follows, four examples of measuring and applying stiffness properties will be presented that deal with issues of homogeneity and isotropy with respect to plastics. In the first case, an amorphous thermoplastic polymer, polycarbonate, is discussed. In this case, the engineering approximations of homogeneity and isotropy are reasonable, and discussion is focused on illustrating the accuracy that can be expected in using standard stiffness measurements to predict the stiffness of a part with more general geometry.

In the second example, the material considered is a thermoplastic structural foam. The significant aspect of this material is its inherent nonhomogeneous morphology. In contrast to polycarbonate, it will be shown that standard stiffness measurements applied to structural foam do not always generally characterize this material in an engineering sense. In this case, a less standard procedure is suggested to provide data more accurate for engineering design analysis.

The third discussion focuses on issues of anisotropic and nonhomogeneous material behavior using a glass-filled injection-molded thermoplastic as an example. Both material measurements and their application to general part stiffness are discussed.

The fourth example is also a nonhomogeneous glass-filled thermoplastic. This material differs from the previously mentioned glass-filled material in two important aspects. First, the glass fibers are significantly longer in this latter case, and second, this sheet material is formed into its structural geometry using the compression molding process. Although this material—often referred to as glass-mat thermoplastic (GMT)—can also display anisotropic stiffness, the focus of the discussion will center on how its nonhomogeneous nature affects the measurement and application of stiffness properties.

These four examples obviously do not represent an exhaustive treatment of stiffness issues in plastics. Rather, the goal is to highlight issues that can be significant in a logical process of mechanical analysis for stiffness and illustrate approaches that have been employed to ensure accuracy in the general prediction of part stiffness.

Part stiffness prediction from standard stress–strain measurements

Perhaps one of the first tasks faced by an analyst investigating the structural response of an engineering component using a new material is identification of the essential material properties necessary to predict quantitatively the stiffness and deformation of the engineering component. Also of great importance is establishing whether or not existing modeling techniques can be accurately utilized with these properties to predict deformation. The basic nature of these questions require that they be addressed early in the development of mechanical technology so that approaches for analysis and design can be built on a firm foundation. In light of the previously identified issues of stress–strain nonlinearity and large displacements and strains that can be expected when dealing with plastic materials, confidence in material characterization, material models, and analytic techniques that will be applied to component analysis is especially important.

The first discussion relative to the predictability of stiffness in plastic parts based upon measured material properties will be a relatively straightforward example involving polycarbonate, which is a homogeneous and isotropic material. We will consider the deformation of a simple open-ended box section shown in Fig. 4.8. The cross section of this box is representative of a geometry considered in early automotive bumper development. It was actually molded in two pieces: a flat backplate and a "c-section," which constituted the front, top, and bottom of the box. These two pieces were then joined using a process known as *vibration welding*. In this process, two pieces of plastic are held together under pressure and then exposed to a high-frequency, oscillatory vibration creating relative motion in the plane of the interface between the two parts. This rubbing process creates high local temperatures at the interface, melting the plastic. When the motion is stopped, the plastic resolidifies at the joint and the two parts are joined. "Flash" that has been squeezed out from the interface is visible at the back lower corners of the box in Fig. 4.8, as well as in the photograph of the test setup in Fig. 4.9. Additional details relative to this joining process can be found in Refs. 3 to 12.

For the test considered here, which is discussed more thoroughly in Refs. 13 and 14, a 10.2 cm (4 in) long piece of this box was cut out and mounted with the backplate of the box clamped between two aluminum plates, as shown in Fig. 4.9. A load was then applied in the form of the uniform displacement of an aluminum bar across the full 10.2 cm (4 in) length of the "nose" on the front face, as shown in Fig. 4.9. During the test, the applied load and displacement under the load were measured.

The material from which the box section was molded is a rubber-toughened, amorphous thermoplastic polycarbonate. At low strain levels, homogeneous and isotropic material response is a reasonable expectation for this material. Before the box section was tested, standard ASTM tensile specimens of the same material grade were molded and tested. Using an extensometer to measure strain, the stress–strain response of this material was measured to yield stress levels and is reported in Fig. 4.10. In general, the stress–strain curves are nearly

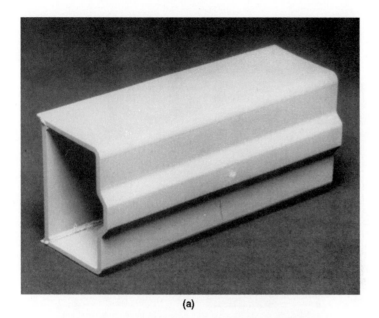

(a)

(b)

Figure 4.8   Hollow box test section.

Figure 4.9 Test configuration for 10.2 cm (4 in) long box-section test.

linear for stress levels less than 21 MPa (3.0×10^3 psi). Thereafter, the stress–strain relationship becomes increasingly nonlinear and reaches an initial maximum stress of about 55 MPa (8.0×10^3 psi) for the relatively slow strain rate of 10^{-3} s^{-1} used in this test. Careful tests of polycarbonate have shown[15] that for stress levels below this maximum stress, the strain incurred during loading is recoverable, although as the applied stress approaches the maximum, some of the strain may be recovered in a time-dependent fashion. In the comparison that follows, two material models will be used to approximate the material behavior and their relative effectiveness in predicting the stiffness of the box section under load will be compared. The simplest material model will be linear-elastic in nature. For this model, a Young's modulus of 2.07 GPa (3.0×10^5 psi) and a Poisson ratio of 0.4 are used to define the elastic deformation. The second material model applied in this example is a multilinear model that more accurately describes the stress–strain behavior between 21 and 55 MPa (3×10^3 and 8×10^3 psi). The discrete points used to establish this multilinear representation of the stress–strain behavior are also shown in Fig. 4.10. Both of these models can be commonly found in commercial finite element codes. For this discussion, we will be primarily interested in the stiffness prediction of the component when stresses are less than 55 MPa (8×10^3 psi). In

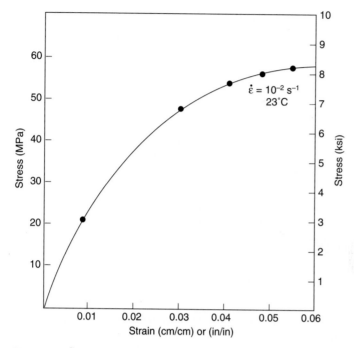

Figure 4.10 Stress–strain curve for rubber-toughened polycarbonate.

Chap. 5, we will reexamine this experiment with emphasis on the prediction of yield phenomena in these materials.

The geometry of the component test considered here was described using the finite-element model shown in Fig. 4.11. Since the component's geometry and load are uniform along the x axis (the axis perpendicular to the plane of Fig. 4.11), a two-dimensional plane-strain analysis was applied. A simplified model with only one element through the thickness of the section was also used with reasonable but slightly less accurate results. The increased number of elements through the thickness improves the accuracy of predicted response in this component by allowing for more accurate integration of the nonlinear stress distribution through the section's thickness at the higher load levels. As has been discussed in Chap. 3, the generally low values of Young's modulus associated with unfilled thermoplastics, like the polycarbonate material considered here, result in relatively large strains and displacements under load. Linear structural analysis, most commonly applied to predict component deformation, assumes that the displacements and surface-area changes that take place during deformation can be neglected in comparison to the original geometric description of the component. In addition, linearized equations that

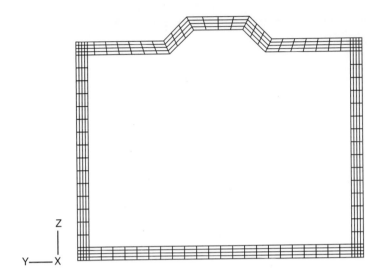

Figure 4.11   Finite-element model of two-dimensional box-section test.

relate the strains in the material to the displacements are also most commonly encountered. As scientific computing capability has progressed, it has become increasingly efficient to relax some of these assumptions when necessary. More and more dependable commercial finite-element codes are now available to include these nonlinear effects in structural calculations. As a result, in the comparisons to be made relative to this example, both linear and geometrically nonlinear finite-element calculations will be offered for comparison with the experimental measurements in order to examine the accuracy of these types of analyses for plastic parts.

Consider the comparison of numerical prediction and experimental measurement of the load–displacement response of this component offered in Fig. 4.12. Three different levels of analysis are presented in Fig. 4.12, all carried out with the same finite-element mesh shown in Fig. 4.11. The linear-elastic (no yielding), small-displacement prediction of the crosshead displacement is accurate to within 20% for loads less than 1.33 kN (300 lb)—only about 37% of the experimentally observed maximum load. If the assumption of small displacements and strains is relaxed by carrying out a geometrically nonlinear analysis, while still modeling the material stress–strain response as linear, the accuracy of the numerical prediction is improved so that less than 20% error is incurred for loads up to approximately 2.66 kN (600 lb). However, if the fully nonlinear stress–strain behavior of the material, as approximated in Fig. 4.10, is included along with the nonlinear strain-displacement equations, extremely accurate (within 5% to 10%) results are attained.

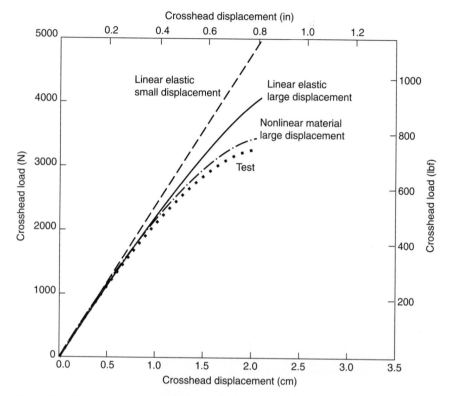

Figure 4.12 Numerical predictions of load–displacement behavior compared to experimental measurement for two-dimensional box test.

This example will be used again to examine capabilities of predicting failure and loading rate effects. However, at this stage it should already be clear from these comparisons that an accurate prediction of a plastic component's stiffness is possible based upon standard stress–strain measurements if the significant nonlinear geometry and material effects are included. Linear analysis has a smaller range of applicability.

Characterizing material stiffness of thermoplastic structural foam

One of the limitations on structural stiffness that can be achieved in parts made of plastics is the maximum wall thickness that can be practically molded. Use of thermoplastic structural foam can sometimes increase the practically attainable wall thickness of a part. The walls of foamed thermoplastic parts have a cellular core with a relatively dense, solid skin. This cellular core, from which the material gets its "foam" name, is achieved either by introducing inert gas directly into the melt or by preblending the resin with a chemical blowing agent. When the resin contacts the mold walls, it forms the relatively dense skin. Later, the gas expands within the plasticized material, producing the internal cellular core. A typical wall cross section of a structural foam part is shown in Fig. 4.13. Since the cavity is only 80% to 90% filled after injection of the material, low mold pressures occur, thus enabling relatively large parts to be fabricated using molds with reduced clamping force requirements. Besides providing approaches to making larger, lighter parts with thicker, stiffer walls, the structural foam process offers several other advantages. In flexure, the material near the center of a panel does not contribute as significantly to structural stiffness as the material at the outside. Because the core of the molded structural foam walls is cellular, the flexural stiffness-to-weight ratio of the part is high. The foam process therefore creates naturally effective flexural members. In addition, the process offers advantages in dimensional stability and tolerance control as well as im-

Figure 4.13 Cross section of a thermoplastic foam part.

proved capability to change wall thicknesses in different portions of the component. The foamed core is also more effective in damping out noise. Finally, tooling for parts made with the structural foam process is generally less expensive. On the downside, using this approach does require working with a foaming agent, and the surface appearance of the parts made with this process is generally not as aesthetically appealing. The surface often has a "swirled" marking that may require painting if surface appearance is important to the finished component.

Unfortunately, the characteristically cellular core of foams that provides the advantage of high flexural stiffness-to-density ratios also poses problems for accurately quantifying the material stiffness of the foam. The most basic issue for this nonhomogeneous class of materials is establishing a useful set of material properties that can be used for predicting the behavior of complex structural parts. Reference 16 summarizes some of the early work performed in this area. More recent efforts to define properties useful in component analyses can be found in Refs. 17 to 23. In order to understand the difference in the nature of measuring the stiffness of a foamed material in comparison to a normal injection-molded material, consider the rectangular bar shown in Fig. 4.14a. Let us assume that the actual pointwise Young's modulus variation is described by $E(-y) = E(y)$. Then, the average tensile modulus, E_{TR}, as determined from a tensile test with a load applied in the z direction will be

$$E_{TR} = \int_0^1 E(\eta) \, d\eta = E_0 \qquad (4.2)$$

where $\eta = 2y/d$ is the nondimensional coordinate in the thickness direction and $E(\eta)$ is the local Young's modulus. The strain ε_z is related to the average stress in the z direction, σ_{average}, through $\sigma_{\text{average}} = E_{TR}\varepsilon_z$.

Next consider the pure bending of this bar in the plate mode (deflection perpendicular to foam layers) shown in Fig. 4.14b. Using standard methods, the average bending modulus, E_{BR}, that would be determined in a flexural experiment is defined by the equation

$$E_{BR} = M_x R / I_x \qquad (4.3)$$

where I_x is the second moment of the beam cross section and R is the radius of curvature of the neutral axis. Assuming a linear distribution of strain through the wall thickness, the apparent bending modulus E_{BR} can be related to $E = E(y)$ through

$$E_{BR} = 3\int_0^1 \eta^2 E(\eta) \, d\eta = E_2 \qquad (4.4)$$

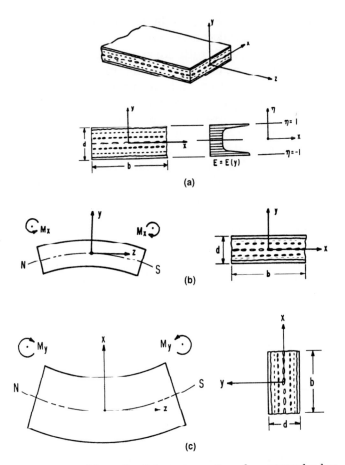

Figure 4.14 (a) Schematic of the cross section of a rectangular bar made of thermoplastic foam. (b) Schematic of a rectangular bar made of thermoplastic foam subjected to bending in the plate mode. (c) Schematic of a rectangular bar made of thermoplastic foam subjected to bending in the in-plane mode.

It should be noted that for a general distribution of local modulus $E(y)$, the expressions defining tensile modulus and flexural modulus in Eqs. (4.2) and (4.4) are clearly different.

Finally, consider the bending of this rectangular bar in the in-plane mode (deflection in the plane of the foam layers) shown in Fig. 4.14c. Following a parallel approach as for Eqs. (4.3) and (4.4), nonhomogeneous beam theory predicts[17–19] that the average flexural modulus for this case equals the tensile modulus in Eq. (4.2).

The important point to be drawn from this discussion is that unlike a homogeneous material, the modulus that is measured during standard tensile tests will not necessarily accurately predict the stiffness of a nonhomogeneous foam plate in flexure. As a result, neither E_{TR} nor E_{BR} alone can be regarded as material properties independent of part geometry. They are averaged quantities that are affected by local, non-homogeneous material morphology. In order for these average moduli to be useful in an engineering sense, procedures must be defined for their application to the analysis of complex foam structures subjected to general loads.

Although a homogeneous material model will clearly have limited use in predicting the stiffness of a general structural part made of structural foam, there are other material models that can be used to measure and predict the stiffness of structural foam more effectively. Because the skin of structural foam, shown in Fig. 4.15a, is generally less porous than the core, the local Young's modulus will be larger in the skin and the variation through the thickness of the part can be approximated by the model shown schematically in Fig. 4.15b. This layered material approximation can then be used in conjunction with laminated finite elements such as the one shown in Figure 4.15c.[21,22] The three parameters that are necessary to specify completely the approximate modulus distribution through the thickness of a foam wall are the moduli of the core and skin layers, E_C and E_S, and the nondimensional thickness of the foam skin, $\eta_0 = 2d_s/d$, where d_s is the skin thickness and d is the total thickness of the foam wall. Using this model, the structural foam can now be approximated as a nonhomogeneous, layered continuum. The integrations described in Eqs. (4.2) and (4.4) for E_{TR} and E_{BR} can be performed using this specific layered

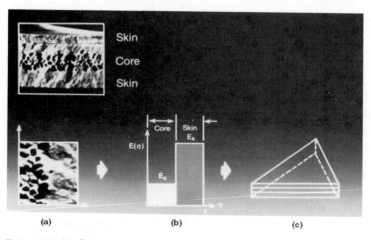

Figure 4.15 (a) Cross section of a foamed wall thickness; (b) approximated variation of the local Young's modulus through the thickness of a part; (c) laminated finite element.

model. Having carried out these integrations, the skin and core moduli E_S and E_C can be determined for any nondimensional skin thickness η_0 by inverting the expressions for E_{TR} and E_{BR} in Eqs. (4.2) and (4.4). Carrying out this inversion results in

$$E_S / E_{TR} = 1 + (m - 1) / [\eta_0(2 - \eta_0)] \qquad (4.5a)$$

$$E_C / E_{TR} = 1 - (m - 1) / [(1 - \eta_0)(2 - \eta_0)] \qquad (4.5b)$$

where $m = E_{BR}/E_{TR}$. Now E_{TR} and E_{BR} can be measured through standard tensile and plate-mode bending tests. For any arbitrarily chosen value of η_0, the material model based upon the simplified variation of moduli shown in Fig. 4.15b and the values of E_C and E_S calculated from Eqs. (4.5a) and (4.5b) will consistently account for the difference in the stiffness measured in bending and tension tests of structural foam. Reference 22 suggests that $\eta_0 = 0.5$ is a practically reasonable value to use. This is a significant advantage since the material model can now be applied to more general geometries where bending and in-plane direct stress are both present. Furthermore, these properties can be used in conjunction with layered finite-element analysis techniques, which are commonly available in commercial finite-element codes.

For effective implementation of this approach to modeling the general stiffness of structural foam parts, a second fundamentally important observation relative to the test specimens used to measure E_{TR} and E_{BR} must be applied. At present, the bulk of the mechanical properties of foams are determined by tests on dog-bone-shaped molded specimens that are specified by ASTM for tensile tests on homogeneous materials. Most foam components are thin platelike structures that have the typical skin–core–skin morphology shown in Fig. 4.13. However, as shown in Fig. 4.16a, the morphology of the one visible in the individually molded dog-bone specimens is quite different than the part cross section in Fig. 4.13 in that the skin layer completely surrounds the core of the dog-bone specimen. Thus, in comparison to common plate or shell structures made of structural foam, individually molded specimens have more skin in relation to the core and are therefore not representative of a typical foam component. Specimens cut from molded plates, like those shown in Fig. 4.16b are therefore recommended in Ref. 20 as more appropriate.

Using the approach suggested in Ref. 20, both flexural and tensile moduli were measured on a structural foam made of a modified polyphenylene ether (M-PPE) material. All the tests were carried out at an ambient temperature of 22°C (72°F) and a 50% relative humidity. Rigid structural foams are described in terms of a general density-reduction level due to foaming. That level of "nominal" density reduction is controlled by the process parameters and is directly related to the

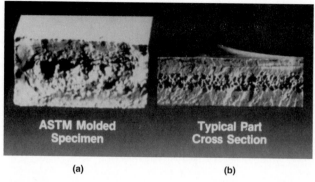

(a)   (b)

Figure 4.16 (a) Morphology of individually molded thermoplastic foam dog-bone cross section; (b) morphology of specimens cut from a foamed thermoplastic plate.

amount of material injected into the mold. The actual density reduction may vary somewhat over the part. In addition to the nominal density reduction of the molded plaques, the data summarized here from Ref. 21 also report the measured average density of each specimen tested.

Results of this testing program for a 5% nominal density reduction M-PPE structural foam are illustrated in Fig. 4.17. In this figure, both

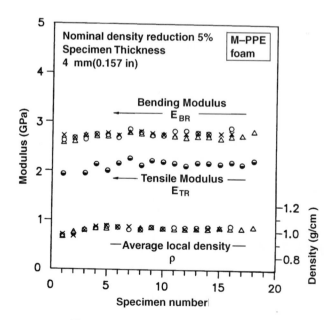

Figure 4.17 Flexural and tensile moduli measured from specimens cut from a thermoplastic foam plate with 5% nominal weight reduction.

flexural and the tensile moduli measured for each of the specimens are reported as a function of specimen number. Specimen 1 was located farthest from the gate end of the plate. Bending tests were carried out for three separate plates of this material and are reported in Fig. 4.17. The data from these three plates of material are remarkably consistent. In each of the three plates, the specimens farthest from the gated edge have lower average densities and elastic moduli. There is also a clear difference between the magnitudes of the flexural and tensile moduli. For the average local densities indicated in the figure, the bending modulus for this material is approximately 30% higher than the tensile modulus.

Since foamed material is used in parts over a range of wall thicknesses, the question of wall-thickness dependence is also important. Figure 4.18 presents tensile and flexural data measured for two different panel thicknesses as a function of specimen number. As can be seen, the measured values for both moduli do not appear to be significantly affected by the plate thickness and can therefore be employed in general.

As one might expect, the density of the foam has a significant effect upon the stiffness of this material. In Fig. 4.19, both nondimensional tensile and bending moduli of the M-PPE foam specimens are plotted

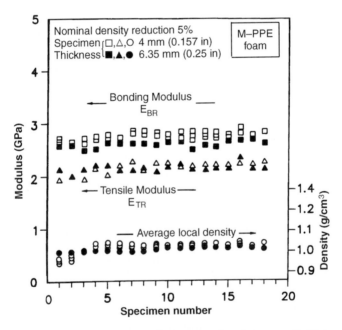

Figure 4.18 Flexural and tensile modulus data for two wall thicknesses of 5% nominal weight reduction M-PPE foam.

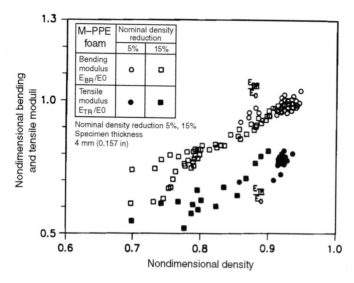

Figure 4.19 Flexural and tensile modulus data as a function of nondimensional density.

as a function of the foam's nondimensional density. In this figure, the modulus E_0 and the density ρ_0 used in the nondimensionalization are the values associated with the unfoamed M-PPE resin. As can be seen, for larger values of average density, both the tensile modulus E_{TR} and the flexural modulus E_{BR} are larger. It is also clear from Fig. 4.19 that there is scatter associated with the measured moduli for any given density, especially for the smallest measured densities. However, the trends are quite strong and the difference between tensile and flexural moduli remain over a wide range of densities. In light of these characteristics, Ref. 23 suggests that data such as those shown in Fig. 4.19 be presented in engineering material databases along with a least-error approximation to the data in terms of density to a power. When used in conjunction with the three-parameter material approximation summarized in Eqs. (4.5a) and (4.5b), this set of data can provide the best information available with regard to the structural stiffness of foamed parts as a function of the foam density reduction.

As an illustration of the improved stiffness prediction available using the suggested testing procedures and the three-parameter, nonhomogeneous, layered material model, two examples from Ref. 23 are offered. In the first case, the part geometry considered was a simple five-sided box loaded with the open end resting on a flat surface and the center of the opposing face loaded with a hemispherical plunger as shown in Fig. 4.20a. The material used to make this box was M-PPE. The initial, linear structural stiffness of the box was measured experi-

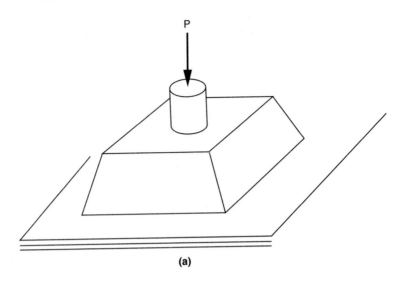

(a)

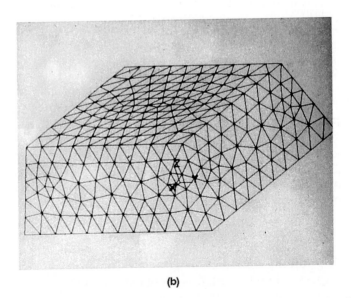

(b)

Figure 4.20 (*a*) Schematic of component test of a five-sided box loaded with a hemispherical indenter; (*b*) finite-element model of component test.

mentally and then compared with three different approaches to predict the stiffness. The linear stiffness predicted using each approach to represent the material behavior is normalized by the experimentally measured stiffness of the part and reported in Table 4.1. A value of 1 for this normalized stiffness would indicate complete agreement between the prediction and the experiment. First, standard flexural modulus data measured on individually molded bars of the material were used in conjunction with a homogeneous, isotropic material model. It will be recalled that in the previous discussion it was pointed out that injection-molded bars of foam are observed to have more skin material than is observed in complex, platelike parts molded from the same material. As can be seen from Table 4.1, when the molded bar flexural modulus is used in conjunction with a homogeneous material model, the finite-element prediction of the part stiffness using the model shown in Fig. 4.20*b* is 110% higher than the experimental result. This is a practical example of how much the stiffness of a structural foam part may be overestimated if test data from injection-molded bars of foam material are applied in structural analyses. Next, the same homogeneous material model and finite-element discretization were used with flexural moduli data measured from flexural specimens that were cut directly from a plate of foam material, as suggested in Ref. 20. As can be seen, the use of these alternate test specimens to measure bending modulus does improve the accuracy of the predicted stiffness, but the results are still 70% higher than the measured stiffness. Finally, both tensile modulus data and flexural modulus data measured from specimens cut from a plate of foam material were applied in conjunction with the three-parameter, layered, nonhomogeneous material model. Using the same finite-element discretization once again, the predicted results reported in Table 4.1 were only 7% higher than the measured stiffness— well within acceptable engineering limits of accuracy.

The second part that was tested and compared with these three approaches to predicting stiffness was a fishing down-rigger molded of polycarbonate foam, illustrated in Fig. 4.21*a*. The finite-element dis-

TABLE 4.1 Comparison of Analytical Predictions and Experimental Results for the Rigid Thermoplastic Foam Box and Fishing Down-Rigger

	Normalized linear stiffness	
	Five-sided box	Fishing down-rigger
Experiment	1.00	1.00
Datasheet ASTM bending data	2.10	1.45
New test technique bending data	1.70	1.05
Skin–core–skin model	1.07	0.94

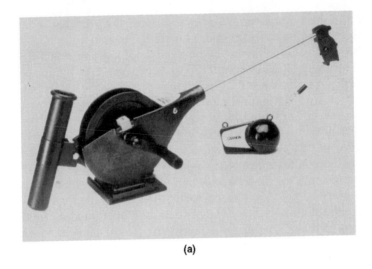

(a)

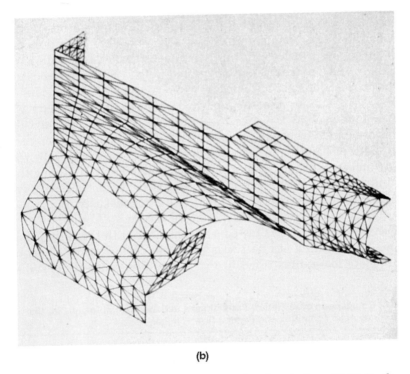

(b)

Figure 4.21 (a) Fishing down-rigger made of polycarbonate foam. (b) Finite-element discretization of fishing down-rigger. (Part symmetry allows only half of the part to be modeled.)

cretization of one-half of this symmetric part geometry is shown in Fig. 4.21b. In this experiment, the base of the part was held clamped while a load was applied to its opposite end in order to closely simulate the part's intended use. The same three predictive approaches were employed and compared in Table 4.1. Again, using the flexural modulus measured on an injection-molded bar of the foam leads to higher predicted structural stiffnesses than those measured in experiment—45% higher in this case. When the bending modulus measured with the specimens cut from a foam plate was used, it yielded much better agreement between the prediction and measurement of the stiffness of this particular component—only a 5% difference. When the complete set of flexural and tensile modulus data are employed using the three-parameter nonhomogeneous material model, the prediction is approximately equal in accuracy to the previous analysis—this time the prediction is about 6% less stiff than the experiment.

In summary, then, the nonhomogeneous nature of structural foam material requires careful treatment both with respect to measurement of material properties as well as with respect to the material model employed to predict the stiffness of parts with general geometry. Because of the characteristic layered, skin–core–skin morphology of the foam, it appears to be more accurate to use material data measured from specimens cut from a plate of foam material instead of using data from individually molded specimens. Furthermore, the so-called flexural modulus measured during such a test is not really a true material property. It will not accurately predict the modulus of a bar from the same foam plate in tension or even in bending when the axis of bending is rotated 90°. The reason for this disagreement is based in the nonhomogeneous nature of the foam material. However, if instead of a homogeneous material model, a nonhomogeneous layered model is employed to represent the foam behavior, then data from both tensile and bending tests on specimens cut from a foam plate can be employed, along with a layered material model available in most commercial finite-element codes, to predict accurately the stiffness of structural foam parts.

In spite of the improved ability to predict structural stiffness of foamed parts evident from these examples, there are clearly still gaps in the existing technology. As shown in Fig. 4.19, the moduli used to define stiffness of a foamed thermoplastic are clearly dependent upon the material's local density. However, this density may vary over the geometry of a part. Although the variation is not significant in the plate of foam with nominal density reduction of 5%, as shown in Fig. 4.18, it can be more significant for larger density reductions or more complex parts. Without an ability to predict the local density of the foam as a function of process parameters and part geometry, the engineer must approximate the material's density and expect an associated error in the ability to predict part stiffness. Nonetheless, the technology of relating material stiffness to part stiffness is clearly capable of better approximations than were available in the recent past.

Material stiffness issues for injection-molded short-glass-fiber-filled thermoplastic materials.   Another approach that is often used to increase the stiffness of thermoplastic-based materials is to fill the plastic with glass fiber. This can be accomplished with a number of different processing approaches. One approach is to combine glass fiber with thermoplastic at the very early stage of producing thermoplastic pellets. In this case the glass fibers used are extremely short and parts can be produced using normal injection-molding techniques. The addition of short glass fibers to the thermoplastic resin can increase the Young's modulus of the filled material to values between 5 and 10 GPa (0.7×10^6 and 1.4×10^6 psi) in comparison to values from 2 to 2.5 GPa (0.3×10^6 and 0.36×10^6 psi) for unfilled thermoplastics. However, the process of adding glass to thermoplastic material also has the effect of creating a nonhomogeneous and anisotropic material and raises several issues relative to engineering design and analysis.

The mechanical performance of glass-filled thermoplastics has been extensively studied and some of the results are reported in Refs. 24 to 29. Many of these studies have focused on the microstructure and behavior of individual fibers surrounded by resin. In general, orientation of these glass fibers during the flow process of injection molding is the fundamental cause for several specific practical issues. The first of these issues is material anisotropy due to the flow-induced orientation of the short fibers. It has been generally observed that the stiffness in the direction perpendicular to the melt-flow direction is significantly lower than that for the flow direction. This behavior is not only a complicating factor for components with complex shape, but can also be a significant issue with respect to the simple specimen geometries commonly used to measure mechanical properties. If typical end-gated, injection-molded dog-bone specimens such as those shown schematically in Fig. 4.22a are used, for example, the narrow geometry induces the majority of the fibers to align in the injection flow direction along the length of the specimen. As a result, when such specimens are used to measure the tensile modulus of this class of material, nonconservatively high values of stiffness are usually measured. Another issue associated with flow-induced fiber orientation is related to the observation that injection-molded, short-glass-fiber-filled thermoplastics can develop layered morphologies with fibers near a mold surface oriented parallel to the injection-molding flow and fibers in the central core ori-

properties are required to completely specify a material within the context of this theory. In terms of so-called "physical constants" these properties are the direct stress moduli in the two principal directions, E_1 and E_2; Poisson ratio v_{12} (which defines the absolute value of the ratio of strain in the 2 direction to the strain in the 1 direction when a unidirectional stress σ is applied in the 1 direction); and the in-plane shear stiffness G_{12}. Assuming that the principal orthotropic moduli are in the flow and cross-flow directions, these four material constants can be measured in three tests. Using tensile specimens cut from a fan-gated plaque, the major stiffness E_1, and Poisson ratio v_{12} can both be measured during a tensile test in the flow direction. The minor stiffness E_2, can be measured during a tensile test in the cross-flow direction. Finally, using these measurements the shear modulus G_{12} can be calculated from measurements of the tensile stiffness of a coupon cut at a 45° angle to the flow direction (E_{45}) using the orthotropic elasticity relationship

$$\frac{1}{G_{12}} = \frac{4}{E_{45}} - \frac{1}{E_1} - \frac{1}{E_2} + \frac{2v_{12}}{E_1} \qquad (4.6)$$

With these properties, the Young's modulus in a direction θ to the flow axis, E_θ, can be calculated for any angle θ and compared with experimental measurements. In order to carry out this comparison, tensile specimens were cut from fan-gated plaques with specimen load axes oriented at angles of 0°, 15°, 30°, 45°, 60°, 75°, and 90° to the flow direction as shown in Fig. 4.25a. Measurements made on the specimens oriented in the 0°, 45°, and 90° directions were used to define the four necessary orthotropic constants. The specimens in the remaining orientations were then used to assess the accuracy of predicting stiffness in an arbitrary direction, θ, using this orthotropic elastic material model. The results of this comparison are presented in Fig. 4.25b for three materials: 20% (by weight) glass-reinforced M-PPE, 30% glass-reinforced M-PPE, and 30% glass-reinforced PBT. As can be seen in the figure, the computed values for stiffness in the various directions based upon orthotropic elasticity agreed well with test values for all three materials.

As a very simple initial assessment of the accuracy associated with applying this material model to more general configurations than a tensile specimen, three-point bending tests were carried out on flat plates, shown in Fig. 4.26a, which were cut from panels made of 30% glass-reinforced M-PPE similar to those used to assemble the necessary material data for the orthotropic elasticity model. Three orientations of the beam length with respect to the flow direction were considered: beam length parallel to flow direction, perpendicular to flow direction, and at an angle of 54° to the flow direction of the panel.

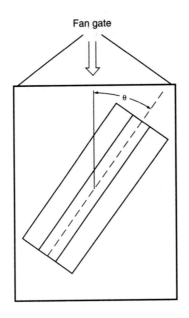

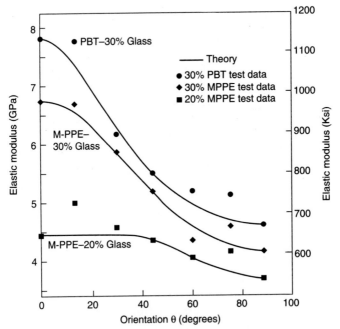

Figure 4.25 Measured and predicted (orthotropic theory) tensile modulus as a function of orientation (θ) to flow direction.

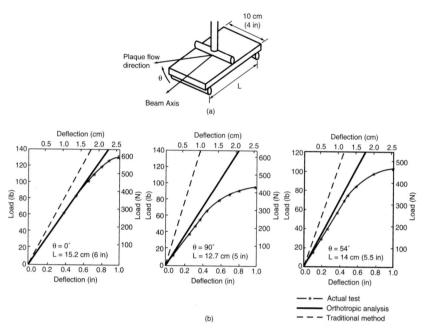

Figure 4.26 (*a*) Schematic of three-point bend verification test. (*b*) Typical measured load–displacement curves for beams made of 30% glass-filled M-PPE compared with predictions based on orthotropic theory using orientation-dependent material data and traditional isotropic assumptions using injection-molded bar data.

The errors between stiffness test predictions and results are summarized in Table 4.3 and examples of the load–displacement curves for each orientation are shown in Fig. 4.26*b*. For each orientation, three curves are presented: the results of the actual test, the predictions based upon the orthotropic elasticity model, and predictions based upon the traditional method using standard stiffness data from individually molded tensile specimens. As a result of geometry constraints associated with the original panel, the beam lengths are different for each orientation and are listed on each graph. Since tensile moduli were measured in the flow and cross-flow directions as part of the data defining the orthotropic model, one might expect the flow and cross-flow bending tests to agree well with predictions. However, the agreement between the tensile and bending tests in these directions does indicate again that significant layering of material, such as was observed in the thermoplastic foams, is not as quantitatively important here. Furthermore, the orthotropic elastic model is very effective in predicting the beam deformation in the 54° specimen. All of the predictions using this model are within 8% of measured results. In comparison, using stiffness data from the individually molded specimens and

TABLE 4.3 Summary of Errors between Linear Stiffness Predictions and Measurements for Three-Point Bending Specimens Made of Short-Glass-Fiber-Filled Injection-Molded Thermoplastic

Material	M-PPE	M-PPE	M-PPE	PBT
Glass filler	20%	30%	30%	30%
Thickness	3.2 mm (0.125 in)	3.2 mm (0.125 in)	5.1 mm (0.200 in)	3.2 mm (0.125 in)
Flow	−5.6	−4.0	+1.2	−7.9
Cross flow	+7.4	+5.1	+5.0	−5.2
54° angle	−3.1	+5.4	...	−0.7

isotropic elasticity can significantly overpredict stiffness, as can be seen in these figures.

Normally, when orthotropic elasticity is applied, the principal material directions are well known to the analyst. In unidirectional composites, for example, the principal directions are parallel and perpendicular to the fibers whose orientation is well defined in space. This is not generally the case for injection-molded parts made of short-glass-fiber reinforced thermoplastics. The previous examples indicate that if the flow direction and cross-flow directions in an injection-molded plate are associated with the principal directions of an orthotropic elasticity model, the stiffnesses in directions at arbitrary angles to the flow can be reasonably approximated both in tension and flexure. However, in a general component, the flow direction will be changing throughout the component and the engineering analyst will not know the flow direction *a priori*. One approach to eliminating this uncertainty is to make use of so-called mold-filling analyses to define the melt-flow directions in the component of interest.

Modeling the mold-filling process in plastic parts involves mechanical analyses relative to fluid flow and heat transfer. Such analyses are currently carried out using commercially available software codes. The molten plastic is usually modeled as a nonisothermal non-Newtonian fluid using thin-shell finite elements to describe the mold cavity that is used to produce the part. One of the parameters that is generally a product of this type of analysis is usually the flow orientation of the plastic melt in each of the elements of the model. Elements used to perform these calculations can be defined in an identical fashion to the elements used to carry out a structural analysis. As a result, a mold-filling analysis can be used to calculate the principal material direction, which by this procedure is assumed to be the melt-flow direction, in every structural element of the model automatically. These principal directions, in conjunction with the orthotropic elastic data measured from specimens cut in three directions from a fan-gated plaque, fully define the material parameters necessary to use the orthotropic elastic model to predict stiffness of a glass-fiber-reinforced plastic part.

As a means of addressing the accuracy of applying this approach to predicting stiffness of a part with more general geometry, a CRT housing, shown schematically in Fig. 4.27a, was chosen for testing. This geometry was molded from 30% glass-reinforced M-PPE using both an edge gate and a center gate, as illustrated in the figure. These two gate locations provide significantly different mold-filling patterns to assess

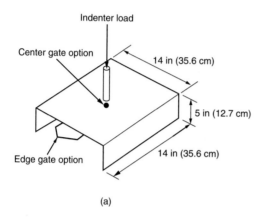

(a)

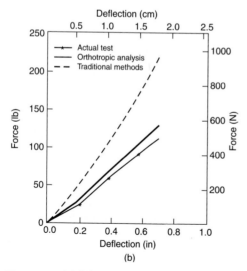

(b)

Figure 4.27 (a) Schematic of CRT-housing test. (b) Typical measured load–displacement curves for an edge-gated housing compared with predictions based on orthotropic theory and traditional isotropic assumptions using injection-molded bar data.

the effectiveness in coupling the melt-flow orientation analysis with the structural stiffness calculation. For this test, the bottom of the three side walls was supported in a fashion to constrain both vertical and horizontal displacement. The centrally located, vertical loading on the top of the housing resulted in large, nonlinear deflections. Using the procedure previously outlined, a mold-filling analysis was carried out for each injection gate location to calculate the melt-flow directions. Under the assumption that these directions represented the spatially varying principal directions of an orthotropic elastic material, the orthotropic properties measured from tensile specimens cut from a fan-gated plaque were used to define the local material stiffness. A non-linear finite-element analysis was then carried out to predict the response of these parts to the applied load. A typical prediction of the displacement under the load for an edge-gate housing is shown in Fig. 4.27b and is compared in that figure to the measured test results. Agreement between test and prediction is quite close. A more traditional prediction based upon individually molded tensile specimen data and an isotropic material model is also shown in Fig. 4.27b and can be seen to significantly overpredict the part's stiffness. Table 4.4 summarizes the accuracy of the two analysis approaches for both edge and center gate locations. The orthotropic predictions were within 10% of measured results for these tests.

Although these results are encouraging, there is need for much more work in this general area. More engineering comparisons of the accuracy of this approach on more complex parts are certainly called for. Such comparisons would help identify the shortcomings and limitations of the approach and focus investigation on areas leading to both improved understanding and analysis techniques.

Material issues for long-glass-fiber compression-molded thermoplastic sheets

An alternative approach to using thermoplastic injection-molding pellets filled with short glass fibers for increased stiffness is to impreg-

TABLE 4.4 Summary of Errors between Linear Stiffness Predictions and Measurements for a CRT Housing Made of Short-Glass-Fiber-Filled Injection-Molded Thermoplastic

	Material	M-PPE	M-PPE	PBT
	Glass filler	20%	30%	30%
Edge gate		−6.1	+8.6	+9.5
Center gate		+6.1	+6.7	+2.5

nate a mat of entangled long glass fibers with a thermoplastic resin, thus producing a composite sheet. The glass fibers in these thermoplastic composites can be either oriented or random. They are usually nonwoven in nature and the fibers are on the order of several centimeters in length. Instead of the pellet form of material used for the short-fiber filled materials, this material is supplied in the form of sheets, which can be used in conjunction with the compression molding process to build up structures of the desired thickness. Heated sheets of this material can undergo a large amount of stretching without glass breakage because the randomly oriented glass uncoils as the deformation increases. Parts can be made from this material by cutting blanks of appropriate size, preheating them in an oven, and stamping the heated material in a press containing both male and female dies. In comparison to the injection-molding process, relatively low pressures are required to form complex parts that have deeply drawn regions, bosses, and ribs. Figure 4.28 is a picture of a bumper beam fabricated with this material and the compression-molding process.

Although this glass-mat thermoplastic (GMT) material provides significant stiffness improvements in comparison to unfilled thermoplastic material—5 GPa (720 psi) for GMT material in comparison to 2 to 2.5 GPa (300 to 360 psi) for unfilled thermoplastics—the nonuniform distribution of glass in the random mat causes the material to be both anisotropic and nonhomogeneous. As a result, elastic moduli measured by standard ASTM tests show an enormous amount of scatter.[30] The

Figure 4.28 Automobile bumper made from GMT composite.

measured moduli also appear to depend on the specimen size.[30] Observations such as these lead to a number of basic questions. Do standard stiffness measurements taken on such a material really reflect a material property? What kind of procedure should be used to identify and apply material measurements to generic design analysis? These issues will be examined in the following section.

A typical GMT thermoplastic composite consists of a 40% (by weight) random glass mat reinforcement in a polypropylene matrix. Figure 4.29 is a radiograph of a 230 × 405 mm (9 × 16 in) plaque of this material in which the random distribution of long glass fibers is very obvious. The glass mat itself, prior to impregnation, is manufactured using a linear process that introduces some directionality with respect to material stiffness properties. Furthermore, when the material is formed during the compression-molding process, the potential difference in size between the molding sheet and the mold cavity can lead to additional directionality introduced as the glass-impregnated sheet flows to

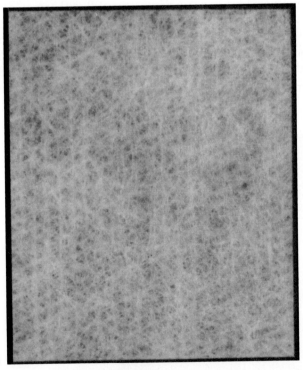

Figure 4.29 Radiograph of a compression-molded plaque of glass-mat thermoplastic composite.

fill the entire mold. Although such effects must be considered in developing a comprehensive design analysis approach for these materials, they will not be emphasized in the following discussion. Instead, the focus here will be on the more basic issue of nonhomogeneity and its effect upon properties useful for design analysis. A great deal of fundamental work relative to this material has been reported[31-34] with the goal of establishing useful engineering material properties relative to design for stiffness, and it will be used to discuss material property and analysis procedures relative to this class of materials.

As discussed generally at the beginning of this chapter, the most standard method of accurately measuring the Young's modulus of a material is to load a tensile specimen and measure the induced strain in the material with an extensometer attached to the gauge area of the specimen, as shown schematically in Fig. 4.30. The length over which the extensometer measures deformation may vary, but 1.25 cm (0.5 in) is a fairly standard length. It is assumed in making this measurement that the material is homogeneous over the length of the specimen so that if the extensometer's location in Fig. 4.30 is changed, the same deformation measurement will be recorded. Using this type of standard test procedure, Ref. 30 reports large amounts of scatter in the measured tensile moduli with data ranging from 2.83 to 9.5 GPa (0.41 $\times 10^6$ to 1.4×10^6 psi). With this amount of variation in a quantity as basic as tensile modulus, a number of very important engineering questions arise relative to its use in engineering calculations.

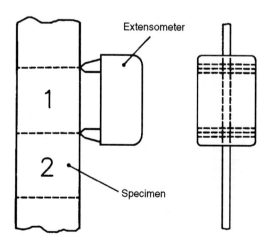

Figure 4.30 Standard single extensometer approach to measuring stiffness of a glass-mat composite.

In order to understand the characteristic variation in the stiffness of this material in more detail, a series of tests on specimens cut from a 230 × 405 mm (9 × 16 in) random-mat plaque shown in Fig. 4.31 were carried out and reported in Ref. 31. For these plates, the 405 mm (16 in) length side of the sheet was aligned with the direction associated with the linear process by which the glass mat is produced. This direction is commonly referred to as the "machine direction." This blank was then heated and thermostamped in a mold of the same size as the blank itself. During this process, there was no "flow" of the glass, but the plaque was subjected to the same thermal and mechanical histories that the material would have to undergo during part forming. This process, in which the material is not made to flow, is called drape molding.

Ten 12.7 mm (0.5 in) bands were marked with a pen on the drape-molded plaque along the machine direction as shown by the dotted lines in Fig. 4.31. Twenty-two 12.7 mm (0.5 in) wide strips were then cut from the plaque parallel to the cross-machine direction as indicated by the solid lines in Fig. 4.31, resulting in twenty-two 12.7 × 230 mm (0.5 × 16 in) specimens with cross-length markers at 12.7 mm (0.5 in) intervals, delineating ten 12.7 mm (0.5 in) segments on each strip. The numbering system for the specimens and the 12.7 mm (0.5 in) segments is shown in Fig. 4.31. For the tensile tests, the specimen numbers increase from 1 at the left to 22 at the right. The 12.7 mm (0.5 in) segments always start with the number 1 at the top and end with the number 10 at the bottom. The mean cross-sectional area of each 12.7

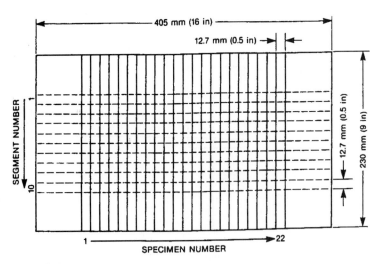

Figure 4.31 Specimen layout in plaque of thermoplastic composite sheet material.

mm (0.5 in) segment on each specimen was determined by measuring the mean thickness and mean width over that 12.7×12.7 mm (0.5×0.5 in) region. The tensile modulus of each specimen at 12.7 mm (0.5 in) intervals was determined via tensile tests, at a strain rate of 10^{-2} s^{-1}, following standardized procedures using an extensometer to measure strain as outlined in Ref. 31. In this way, the Young's modulus parallel to the cross-machine direction of a 127×280 mm (5×11 in) portion of the plaque was determined at every 12.7×12.7 mm (0.5×0.5 in) square unit.

The most graphic way to display the results of these tensile tests is in the form of contour plots of equal-modulus values over the plaque section where measurements were taken. Contour plots of measured modulus variation are shown in Fig. 4.32 based on extensometer measurements made on the right and left sides of the specimen, as well as the averaged values. Detailed examination of the data reveals extremely large differences in the measured tensile modulus, depending not only upon the vertical location of the extensometer on the tensile specimen but also upon whether the extensometer was on the left or right side of the strip. The lowest local modulus measured on this plaque was 2.83 GPa (0.41×10^6 psi) and the highest was 8.78 GPa (1.28×10^6 psi), a ratio of 3.1.

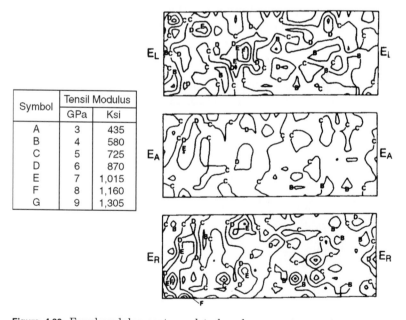

Symbol	Tensil Modulus	
	GPa	Ksi
A	3	435
B	4	580
C	5	725
D	6	870
E	7	1,015
F	8	1,160
G	9	1,305

Figure 4.32 Equal-modulus contour plots based upon extensometer measurements from the left (E_L) and right (E_R) sides of the specimen as well as the average value (E_A).

After the modulus measurements were taken, each specimen was cut into the ten 12.7 mm (0.5 in) square coupons over which the strain measurements had been made. Local densities of each coupon were calculated by measuring coupon mass and dimensions. Use of these data in conjunction with the modulus measurements provides insight into the cause of the fluctuations in measured moduli. Since the moduli measured on each coupon varied depending on whether the extensometer was on the left or right side, the average of the two measurements, E_A, was used to define the modulus of each coupon. Figure 4.33 displays this average modulus, E_A, of each coupon as a function of the coupon's density. As can be seen, there is a strong correlation between the two properties, which indicates that the modulus is closely related to the local glass content in the plaque.

The results of these careful measurements clearly establish some of the character of this material. Quite obviously, these materials exhibit property variations on a macroscale such that the modulus, as customarily measured, can vary by a factor of 3 over a 127×230 mm (5.0×11.0 in) plate. In fact, variations of a factor of 2 can be observed over distances of as little as 12.7 mm (0.5 in). The variations appear to be random. In metals, this type of variation in properties can be expected to occur at the microstructure level. In a parallel sense, these random-mat materials can be said to have a macrostructure over which their properties are not uniform. In addition, this work makes it clear that the variation in the tensile modulus reported from standard ASTM ten-

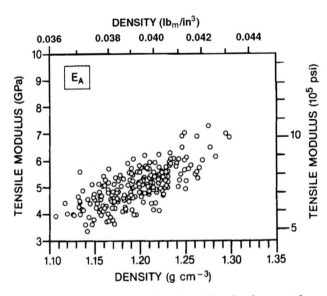

Figure 4.33 Measured average local modulus of a glass-mat thermoplastic composite as a function of local density.

sile tests, such as those reported in Ref. 30, is rooted in this macro-structural morphology of the material. Indeed, it should not only be expected that tensile modulus measurements should vary significantly between tensile specimens, but that within one specimen, the modulus will be measured with substantial variation based simply on where the extensometer is mounted.

When the value of a parameter varies over a significant range in a random fashion, such as is the case for the measured moduli of these thermoplastic composite sheet products, the mathematical concept of a probability density function can be useful in characterizing the expected modulus. It has been shown[34] that this approach is useful in the case of the measured moduli of GMT materials. The probability density function for a discrete set of n modulus measurements can be approximated in a histogram fashion by dividing the range of modulus values into equal segments (ΔE) and plotting the proportion of the total number measurements within each segment divided by the modulus increment ($n_i/n \ \Delta E$). The ordinate of such a plot represents the probability distribution density for the set of measurements. The probability of measuring a modulus between any two values is the integral of the probability density between the two values. An example of such a representation of the average modulus data for a GMT plaque is shown in Fig. 4.34, along with the functional approximation to the data. The de-

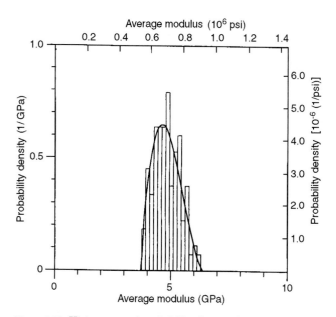

Figure 4.34 Histogram and probability density functions showing experimentally determined distribution of average local modulus.

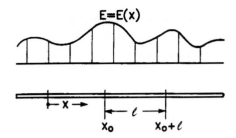

Figure 4.35 Hypothetical variation of elastic modulus of a thermoplastic composite sheet material along the length of a tensile specimen.

gree of scatter in the data can be quantified in terms of the standard deviation of the probability density function and defined for a discrete set of values as

$$\sigma = \sqrt{\sum_{i=1}^{n}(E_i - E_{\text{avg}})^2 \left(\frac{n_i}{n}\right)}$$ (4.7)

The larger the value of σ, the more scatter there is in the set of modulus measurements. Large scatter in the values of measured material moduli is generally considered undesirable from a standpoint of quality, repeatability, and predictability of behavior.

In light of the significant variation in the stiffness of this material on a local level, due to the material's morphology, it is pertinent to question what actually is measured with a standard stiffness test and what relationship it actually has to the performance of more general parts. To begin to answer these questions, it is worth considering the relationship between the measured modulus and the gauge length over which that measurement is made. Some understanding of this relationship can be gained by considering a simple one-dimensional bar made of a model material in which the local elastic modulus, $E(x)$, is assumed to vary continuously with distance along its length, as shown in Fig. 4.35. Then, for this one-dimensional geometry, the strain ε_x along the x direction is related to the stress σ_x through

$$\varepsilon_x = \frac{du}{dx} = \frac{\sigma_x}{E(x)}$$ (4.8)

where u is the x direction displacement. It follows that the extension of the bar between $x = x_0$ and $x = x_0 + l$ is given by

$$\Delta u = u(x_0 + l) - u(x_0) = \int_{x_0}^{x_0+l} du = \int_{x_0}^{x_0+l} \frac{\sigma_x}{E(x)} \, dx$$ (4.9)

iting value. In Ref. 32 it is shown that for the model material defined by Eq. (4.13), this limiting value can be expressed as

$$\left.\frac{E_T}{E_0}\right|_{n\to\infty} = \sqrt{1 - e_0^2} \qquad (4.16)$$

This rather simple element of understanding relative to the measured stiffness of a strip of material with varying modulus provides an opportunity to illustrate two very important generalizations with respect to these materials. The strain measurements used to create the contour plots of stiffness in Fig. 4.32 were all performed with 12.7 mm (0.5 in) gauge lengths. These measurements were in turn used to define the probability density function representing that set of data and shown in Fig. 4.34. The use of a 12.7 mm (0.5 in) gauge length for these measurements is arbitrary in nature. Any gauge length might have been used, and use of the model material indicates that an entirely different set of measured stiffnesses would have been defined. Having discussed the relationship between true local modulus and the modulus measured over a finite gauge length, Eq. (4.12) can be used to create a new set of modulus data that would have been measured over gauge lengths that are multiples of the original 12.7 mm (0.5 in) length. Figure 4.38 illustrates the probability density functions repre-

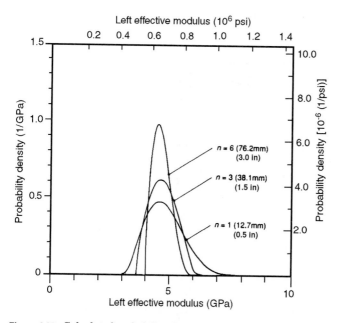

Figure 4.38 Calculated probability density distributions of the measured modulus of a glass-mat-reinforced thermoplastic composite for three gauge lengths [12.7 mm (0.5 in), 38.1 mm (1.5 in), 76.2 mm (3.0 in)].

senting the original experimental measurements made over 12.7 mm (0.5 in) gauge lengths as well as two "simulated" populations of moduli based upon gauge lengths of 38.0 mm (1.5 in) and 76.2 mm (3.0 in). The most noticeable difference in these three functions is that the width, which is related to standard deviation of the measured values, decreases as the gauge length increases.

Figures 4.39 and 4.40 clearly illustrate the two important effects associated with larger gauge length measurements. In Fig. 4.39, the "width" of the probability density function representing the data for a range of gauge lengths is quantified in terms of its standard deviation σ and plotted as a function of the gauge length. Clearly the standard deviation and hence the scatter in measured data decreases as the gauge length increases. Furthermore, Fig. 4.40 illustrates the dependence of the mean (or expected value) of the probability density function representing data associated with a range of gauge lengths and clearly illustrates that as the gauge length increases, the mean of the probability densities asymptotically approaches the harmonic mean, defined in Eqs. (4.11) and (4.12), of the original set of data measured using 12.7 mm (0.5 in) gauge length. The important implications of this information are that over sufficiently large gauge lengths, the measured scatter in stiffness data for this material is significantly reduced.

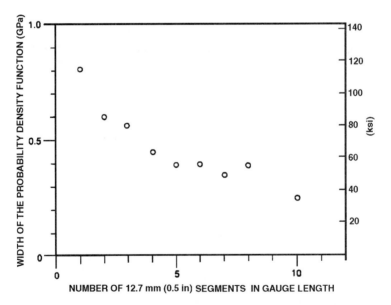

Figure 4.39 Variation of the width of the probability density function with gauge length.

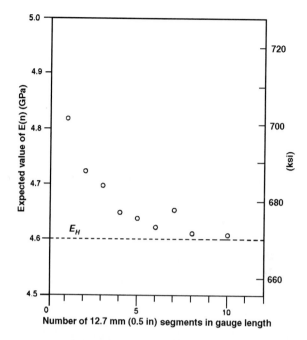

Figure 4.40 Evolution of the expected value of *E(n)* with increasing gauge length.

Furthermore, as was observed for the sinusoidally varying model material, the expected value of the modulus associated with a large gauge length is simply the harmonic mean of the set of data measured using standard 12.7 mm (0.5 in) gauge sections for strain measurement.

Although the previous discussion contributes to an understanding that the harmonic mean modulus of this class of materials measured over a sufficiently large distance may provide a modulus measure for this material that does not show the significant standard deviations characteristic of more local measurements, there is still the important question of whether such a modulus is useful to predict the stiffness of a structure subject to loads other than simple uniform tension. Some initial insight into this issue is also provided in Ref. 32. Here the apparent stiffnesses of a tension member and a simply supported beam under a concentrated central load are considered and compared. For the case of the beam of length *l,* as shown in Fig. 4.41, let the deflection in the *y* direction be denoted as *v.* Then, for small deflections, the displacement of the beam is governed by the Euler–Bernoulli relation:

$$E(x)I(x)\frac{d^2v}{dx^2} = M_x \qquad (4.17)$$

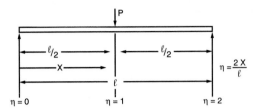

Figure 4.41 Geometry for the bending of a nonhomogeneous beam in a three-point flexural test.

where $E(x)$ is the tensile modulus (tension and compression) at x, $I(x)$ is the second moment of area, and M the bending moment. Let $R(x) = E(x)I(x)$. Then

$$\frac{d^2 v}{dx^2} = \frac{M_x}{R(x)} \tag{4.18}$$

For an arbitrarily varying modulus and second moment of inertia, Ref. 32 shows that the deflection of the beam at its center can be written

$$\delta = \frac{P}{4} \left\{ \frac{l}{2} \int_0^{l/2} \eta \left[\frac{1}{R(\eta)} + \frac{1}{R(l-\eta)} \right] d\eta \right\} - \int_0^{l/2} \int_0^{\xi} \eta \left[\frac{1}{R(\eta)} + \frac{1}{R(l-\eta)} \right] d\eta \, d\xi \tag{4.19}$$

Now, for bend tests on rectangular beams, I is a constant. Furthermore, if an effective flexural modulus E_B is defined by the homogeneous beam formula $\delta = Pl^3/48IE_B$, then the effective flexural modulus E_B can be represented as

$$\frac{1}{E_B} = \frac{12}{l^3} \int_0^{l/2} x^2 \left[\frac{1}{E(x)} + \frac{1}{E(l-x)} \right] dx \tag{4.20}$$

At first sight this expression for the effective bending modulus would appear to be of an entirely different form than that defining the effective tensile modulus given in Eq. (4.11).

Using the model material defined in Eq. (4.13), the same type of calculation can be carried out for the effective bending modulus of the simply supported beam, E_B, as was previously done for the effective tensile modulus. In this case, the length of the beam replaces the length of the gauge over which the tensile modulus is measured. The integration for Eq. (4.20) for the assumed local modulus variation given in Eq. (4.13) is somewhat more complex, and the details of that process are given in Ref. 32. However, in a fashion entirely similar to the effective tensile modulus, the effective bending modulus can be represented as a function of the length of the beam nondimensionalized by the material wavelength λ. This functional dependence is illustrated in Fig. 4.42 for three different values of the material model parameter e_0. Again, for small values of the beam length l, the effective flexural modulus of the model material is dependent upon the beam length. For these smaller values of nondimensional beam length, the effective bending modulus is different than the effective tensile modulus over a gauge length equal to the beam length. However, for beam lengths much larger than the material wavelength, the effective flexural modulus also approaches a limiting value. Most importantly, this limiting value is the same value as that for the effective tensile modulus given by Eq. (4.16). This limiting value of the effective bending modulus is also shown in Fig. 4.42 for three values of the material model parameter e_0.

In conclusion, the large variations in the tensile moduli of random-glass-mat composites should be viewed as a characteristic property of these materials and should not be looked upon as scatter in the sense of deviation from controlled quality.[34] The variation observed for meas-

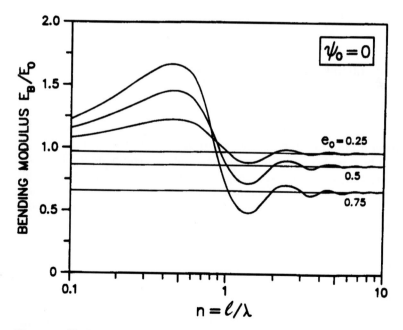

Figure 4.42 Variation of the effective bending modulus hypothetically measured from a three-point flexural test as a function of $n = l/\lambda$.

urements made with standard 12.7 mm (0.5 in) gauge lengths is a reflection of the macrostructure of the material. Careful consideration of the stress fluctuations in long strips indicates that GMT behaves like homogeneous materials in sufficiently large parts. For such parts, the harmonic mean modulus E_H defined in Eq. (4.12) as E_T over the elastic modulus population has been shown to best characterize the large-scale stiffness of these materials. The structural stiffness of large-scale parts made of these materials will be characterized by significantly less scatter than is associated with standard 12.7 mm (0.5 in) gauge length tensile tests.

As is the case of structural foam, it is important to recognize the elements of technology that are still missing with respect to accurate engineering characterization of this class of materials. As was mentioned in this discussion, the compression-molding process used to fabricate GMT parts can lead to directionally dependent flow and consequent directionality in the material elastic moduli. This directionality will certainly be dependent on part geometry. However, to date, there is no approach available to predict this directionality accurately. Furthermore, exhaustive parameter studies defining the relationship between directionality and elastic moduli also do not exist. The development of technology to properly define these characteristics would make the engineering analysis and design process involving these materials still more efficient.

Structural Stiffening Concepts for Plastic Parts

Stiff plastic beams

Since the moduli of engineering plastics are typically $\frac{1}{10}$ to $\frac{1}{30}$ the modulus of aluminum and $\frac{1}{30}$ to $\frac{1}{100}$ that of steel, structural stiffness may often be a real concern in designing a plastic component. However, low material stiffness can sometimes be offset in a structure through the effective use of geometry. Consider, for example, a simply supported beam with a solid, rectangular cross section subjected to a concentrated central load. The deflection of such a beam can be expressed as

$$\delta = \frac{PL^3}{48EI} \tag{4.21}$$

Its stiffness is then expressed as

$$S = \frac{P}{\delta} = \frac{48EI}{L^3} \tag{4.22}$$

where I is the second moment of area of the cross section, defined as

$$I = \int_A z^2 \, dA = \frac{1}{12} b t^3 \qquad (4.23)$$

where b is the width of the beam and t is the depth or thickness. If we insist that the length of the beam remain constant, then a beam made of plastic with a modulus of 2.3 GPa (0.33×10^6 psi) can be made as stiff as a beam made of aluminum with a modulus of 69 GPa (10.0×10^6 psi) either by making the plastic beam 30 times as wide as the aluminum beam or 3 times as thick, the latter being a very dramatic compensation for low material stiffness.

Although increasing the thickness of a solid section beam can compensate for structural stiffness in some situations, there are practical processing limitations with respect to how thick a plastic wall thickness can be manufactured. In situations where still more stiffness is required, thin-walled beams with large second moments of cross-sectional area, such as the box beam in Fig. 4.43, must be considered. Cross-sectional geometry like this allows for even greater improvement in structural stiffness within the limits of thin-walled processing constraints. As we saw in Chap. 2, there are a variety of manufacturing methods available for plastic materials, including injection molding, blow molding, and extrusion, which make fabrication of such cross sections feasible and economical. However, when such a solution is chosen to attain stiffness in a component, there are additional design considerations that must be addressed as part of an effective plastic part design.

Although simple beam theory indicates that increasing the second moment of cross-sectional area of the box beam produces dramatic improvements in stiffness, the plastic beam begins to behave more and more like a thin shell structure than a simple beam. One of the possible by-products of this behavior is that stiffness calculations based upon Euler–Bernoulli beam theory or even general linear structural analysis performed with a finite-element package may become quite inaccurate when used for predicting the stiffness of the beam. Two specific

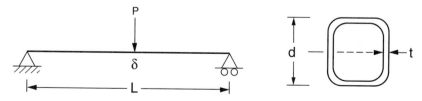

Figure 4.43 Thin-walled thermoplastic box beam.

examples will be discussed relative to this subject. The first example involves some very general issues of boundary conditions for simple beam theory and their relative accuracy in predicting the stiffness of a built-up, thin-walled thermoplastic beam. As part of this discussion, some simple comparisons of plastic beam attachment techniques will be presented and compared on the basis of efficiency in producing stiffness. The second major subject of discussion will be the accuracy of linear analysis techniques in predicting the stiffness of thin-walled thermoplastic beams. Results of nonlinear analysis and linear analysis will be compared and explanations for the differences will be discussed.

Let us consider a thermoplastic box beam (as in Ref. 13), attached at its backplate to rigid supports at two locations, as shown in Fig. 4.44. Effective attachment at these locations can eliminate all lateral dis-

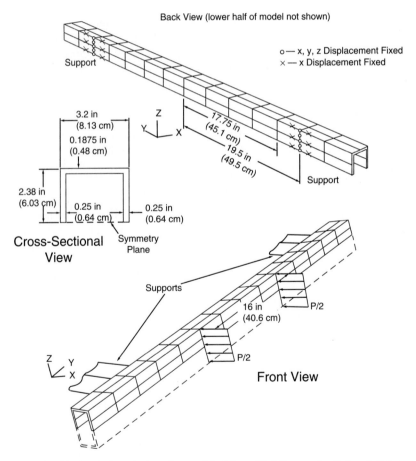

Figure 4.44 Geometric description of a model, thin-walled thermoplastic box beam and its cross section.

placement in the rigid support area. For this discussion, the load will be applied along the full width of the beam with half of the total load P at each of two locations on the front face, as shown in Fig. 4.44. The beam is assumed to be symmetric about the $z = 0$ plane and the finite-element mesh also shown in that figure makes use of that symmetry and represents only the upper half of the box beam. A logical first approach to predicting the stiffness of this beam is to apply simple Euler–Bernoulli beam theory based upon the assumptions that plane sections remain plane and perpendicular to the original centroidal axis of the beam. Using this approach to predicting stiffness requires that the boundary conditions at the support areas be defined within the context of beam theory. That is, the displacements and the rotations at the support areas of the beam can be specified to be either constrained or free. If the support areas are rigid, it is reasonable to expect zero displacement for the beam at this location. However, the boundary condition defining the rotation of the thermoplastic box beam at the support locations is less obvious. Although the substantial support area might seem to be consistent with the idea of restrained rotation at these supports, it is not at all clear that these rotations are realistically constrained to be zero, especially since the physical supports really only affect the back plate of the beam. The front plate is completely unrestrained.

Figure 4.45 compares several different predictions of the load versus backplate deflection at the midlength of the thermoplastic box beam shown in Fig. 4.44. For the comparisons in Fig. 4.45, linear-elastic material behavior is assumed to apply with a Young's modulus of 2.07 GPa (0.3×10^6 psi). Simple solutions based upon Euler–Bernoulli beam theory are compared with both linear and nonlinear finite-element analyses. Two beam-theory solutions are presented in Fig. 4.45. One solution is based on clamped boundary conditions at both ends of the beam—that is, both displacement and rotation of the beam are zero at the support areas. Although the actual plastic beam may have a condition of zero rotation enforced at the backplate of the beam, this boundary condition obviously leads to results that are significantly stiffer than those of the more realistically formulated finite-element models. Alternatively, the boundary conditions of the beam could also be approximated as being simply supported in nature—that is, free to rotate at the support areas, but constrained in lateral displacement. As might be expected, this boundary condition leads to more flexible results than the Euler–Bernoulli beam analysis with clamped conditions. The simply supported solution is also more flexible than the linear finite-element model of the box beam, which more accurately models the beam's boundary conditions as they are realistically applied to the backplate. As can be seen from the comparison of linear analyses in Fig. 4.45, the realistic boundary conditions for this thermoplastic beam cannot be accurately modeled using only Euler–Bernoulli beam theory. The more

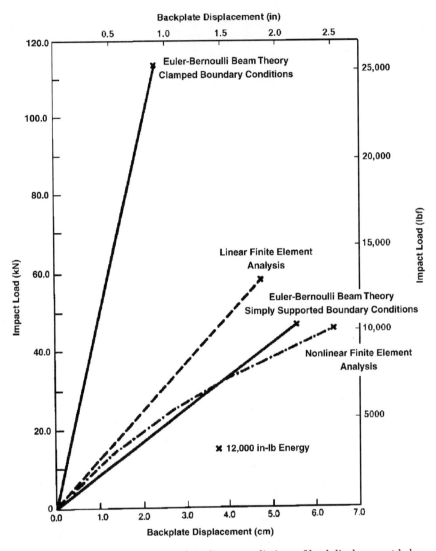

Figure 4.45 Comparison of linear and nonlinear predictions of load displacement behavior for a thin-walled thermoplastic box beam.

detailed definition of the boundary conditions available through use of the finite-element technique is necessary for accuracy in prediction of the stiffness of thin-walled beams with large second moments of area. It is also quite obvious from the nonlinear solution in Fig. 4.45 that there are significant nonlinear effects present in this beam's behavior. These effects will be discussed subsequently.

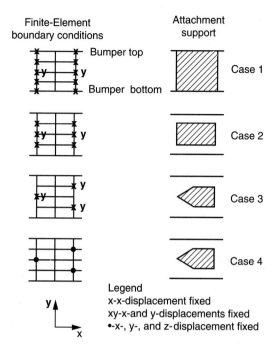

Figure 4.46 Box-beam support conditions and modeled boundary conditions.

Before discussing nonlinear effects, some additional discussion of boundary conditions is warranted. There are a number of possible variations on the support conditions in addition to the ones shown in Fig. 4.44. Several possible variations of the back-plate support are shown in Fig. 4.46, along with the discretely applied constraints used to represent these physical supports with the finite-element analysis method. The relative effects that these boundary conditions have on the beam stiffness are illustrated in Fig. 4.47. As can be seen, in addition to the question of clamped or simply supported conditions, the area over which the attachment takes place has a very significant effect upon the deformation of the thermoplastic beam. Smaller attachment areas at the backplate lead to much larger total displacements of the beam, largely due to local deformation of the backplate in the support area. As a result, the stiffnesses of the beams with support areas illustrated in cases 2 and 3 are significantly lower than that in case 1, which extends over the entire backplate and bears transverse shear load directly from the top and bottom webs of the box section.

Returning now to the nonlinear finite-element solution for the thin-walled box beam with the original boundary conditions in Fig. 4.45, it

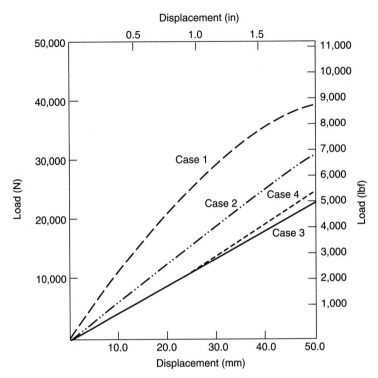

Figure 4.47 Effect of boundary condition variations on box-beam load–deflection behavior.

can be seen that for higher loads, there is a significant difference between the linear and nonlinear results. For loads above 40 kN (9100 lbf), the backplate deflection of the bumper beam predicted with nonlinear analysis techniques is 40% larger than that predicted with linear finite-element techniques. It should be recalled that the material model applied here is linear elastic in nature. As a result, the nonlinear load–deflection behavior visible in Fig. 4.45 is a result of geometric nonlinearity not material nonlinearity. The relatively low modulus of thermoplastics plays a significant role in the importance of the nonlinear effects visible in Fig. 4.45.

As pointed out earlier, the advantage of the box-beam geometry is based on the high second moment of area (I) that is achieved by creating a large distance between the front and back face of the beam cross section. However, under increasing bending loads, the cross section of a thermoplastic box beam will change shape. The front and back face of the box beam move closer together as shown in Fig. 4.48. When this happens, the second moment of cross-sectional area decreases in size

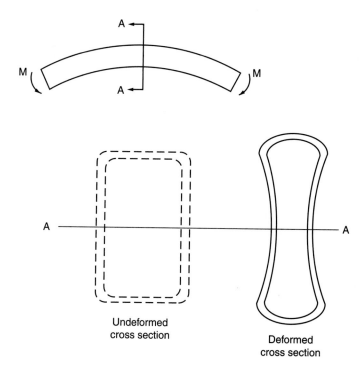

Undeformed
cross section

Deformed
cross section

Figure 4.48 Cross-sectional geometry change of a thin-walled box beam in bending.

and the beam becomes successively less stiff. Even a detailed finite element analysis of this beam will not account for this behavior if it is a linear analysis. Such an analysis is based upon the original geometry of the structure and the assumption that changes in that geometry will be negligible in size. Clearly, this will not always be the case for a thermoplastic box beam. Nonlinear analysis techniques, which are available in many commercial finite-element programs, may be required to define stiffness of thin-walled thermoplastic beams. Although such analyses require more analysis experience than linear techniques, they are being effectively applied.[13,35]

Stiff plastic plates

In a fashion quite parallel to that associated with our discussion of beam stiffness, it is possible to compensate for the low material stiffness of a plastic in a panel configuration by increasing the panel stiffness through effective use of geometry. Of course, just as in the case of a plastic beam, there are practical limitations to this solution for plates also. If panel thickness cannot be increased, one of the most frequently used methods to increase the stiffness of a plastic panel is to add stiffeners. This addition is quite compatible with the injection-molding process so often used to manufacture plastic components. In fact, it is often much easier to fabricate a plastic panel with stiffeners than a panel made of other engineering materials because no secondary process—e.g., gluing, welding, riveting, or machining—is usually required to attach the stiffeners.

Stiffening panels against lateral load has long been the object of engineering studies in the aerospace industry. However, as previously discussed, the large yield strain of thermoplastics makes the geometrically nonlinear range of plate behavior of more practical interest because very large deflections and rotations can be reached without yielding. In the discussion that follows, comparisons in stiffness are made over the entire range of behavior, both linear and nonlinear.

In Ref. 36, the ADINA® nonlinear finite-element code[37] was used to assess the effect of including ribs on injection-molded panels. Four-noded shell elements were employed and Fig. 4.49 illustrates some representative meshes for the analyses performed. In all cases, the flat plate portion of the structure was 1 m (39 in) on a side and 6 mm (0.24

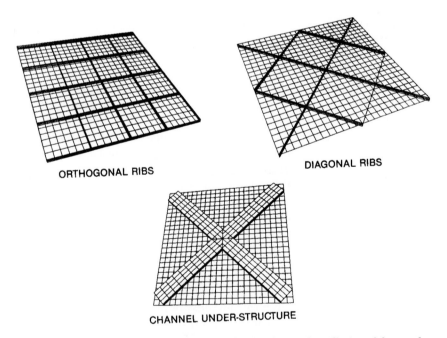

ORTHOGONAL RIBS

DIAGONAL RIBS

CHANNEL UNDER-STRUCTURE

Figure 4.49 Representative finite-element meshes used to study stiffening of thermoplastic panels.

in) thick. Simply supported boundary conditions were applied to the edges of the plate to define the lateral displacements and rotations there. Since geometrically nonlinear analyses were performed, it is also necessary to define the boundary conditions in the plane of the panel. For the analyses discussed here, the in-plane boundary displacements were constrained to be uniform in a direction perpendicular to the free edge.

For the stiffener comparisons presented here, properties representative of a polycarbonate–polybutylene terephthalate thermoplastic blend were used, the Young's modulus of this material is 2.1 GPa (0.3×10^6 psi) and Poisson's ratio is 0.4. Characteristic of the behavior of a thermoplastic panel, the stress response over the entire range of loads reported in this investigation was below yield stress everywhere. As a result, linear-elastic material behavior was assumed.

Figure 4.50 compares the performance of several different rib patterns on predicted stiffness under a uniformly distributed pressure load. Only rib patterns with various numbers of ribs orthogonal to the edges of the panel are compared. During this comparison, the rib thicknesses were held constant at 6 mm (0.24 in) and the rib depths were held constant at 25 mm (1.0 in). There are several results to be noted. First of all, the most effective location to place the ribs is near the cen-

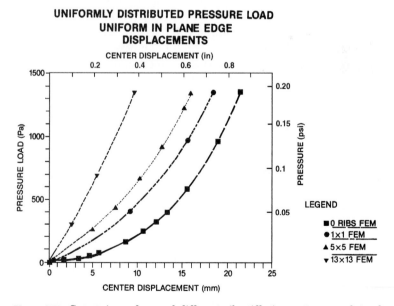

Figure 4.50 Comparison of several different rib stiffening patterns on lateral panel stiffness.

ter of the plate. Placing one pair of ribs orthogonal to each other across the center of the plate reduces the displacement under 1400 Pa (0.2 psi) pressure by almost 20%. In addition, it can be noted that as more highly stiffened panels are investigated (the 13×13 pattern of orthogonal ribs, for example) the results are much more linear in nature. In highly stiffened panels, linear stiffness analyses may be quite sufficient. However, for less heavily ribbed panels, nonlinear results may be significantly stiffer than linear results indicate.

Stiff blow-molded panels

There are other approaches to increasing the structural stiffness of thermoplastic panels in addition to adding ribs to injection-molded panels. The blow-molding process is another manufacturing option that can be applied to make stiff plastic panels by creating double-walled geometries. This concept is the two-dimensional extension of the same idea that makes the box section an efficient geometry for beams—that is, increasing the second moment of cross-sectional area by placing material at greater distances from the neutral axis of the panel.

It will be recalled from an earlier discussion in Chap. 2 that engineered blow molding is a process for producing hollow, double-walled structural parts with engineering thermoplastics. In this process, plastic pellets are loaded into a hopper, melted, and then extruded through a die to form a hollow tube of plastic called a parison. This process is illustrated in Fig. 4.51. After forming, the parison, still at a relatively high temperature, is dropped between two molds, which are then closed. Air is next blown into the parison of molten plastic, forcing it to expand outward onto the cooler inside surfaces of the mold, which forms the shape of the part. After cooling, the molds separate and the part is ejected.

In order for this double-walled, blow-molded approach to be truly effective in producing a stiff panel, however, there must be some means of carrying shear from one face of the panel to the other. If this shear transfer is not provided, the two panel face sheets will simply act as separate, thin plates, greatly reducing the potential stiffness of the panel. The box beam accomplishes this load transfer through the two webs that join the top and bottom face sheets. Two approaches that can be envisioned to provide this shear transfer in blow-molded panels are foam filling the cavity between the two face sheets with a low-density foamed plastic or by periodically creating points of contact (tack-offs) between the two faces through the use of intruding dies while the plastic is still pliable at high temperature. Several examples of many potential geometries that can be created with this latter approach are shown in Fig. 4.52.

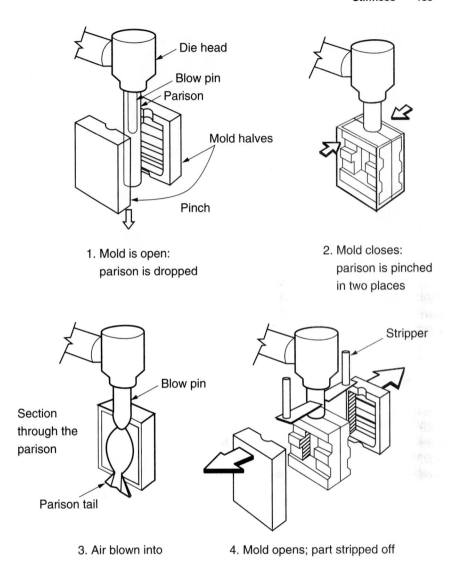

Figure 4.51 Schematic of blow-molding process.

These concepts represent only two of many potential approaches to producing stiff thermoplastic panels through innovative application of plastic materials and their fabrication processes. Engineering analysis can play significant roles in optimizing concepts such as these as well as generating new ideas. There are a number of significant analysis issues that must be addressed to make this possible, however, and several of these will be explored in this section based upon the work pre-

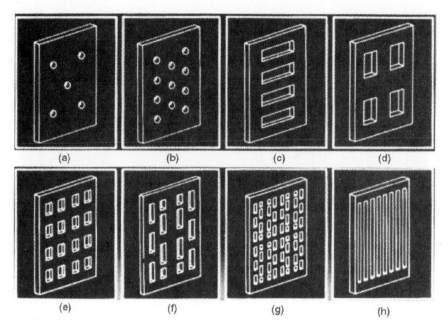

Figure 4.52 Concept of panel stiffening with tack-offs created via the blow-molding process.

sented in Refs. 38–41. First, as in a number of the previous examples discussed in this chapter, material properties must be appropriately defined so that the analyses can be carried out with confidence. Having established these properties, comparisons with experiments can be used to assess the fundamental accuracy of the predictions. From this position of confidence, parameter studies can be carried out to quantify the macroscopic improvements that are possible by changing geometric and material properties. All of these issues will be considered within the following discussion focused upon optimizing stiffness in plastic panel structures.

The thermoplastic that was used in the blow molding of the panels in this investigation was a blow-molding grade of M-PPE. In the case of the foam-filled panels, the foam core was a rigid, low-density urethane foam. A 31 N (7 lbf) Sterling blow-molding machine was used to fabricate blow-molded M-PPE panels that were 63.5 cm (25 in) long and 25.4 cm (10 in) wide. The depth of the panels from outside face surface to outside face surface was 1.9 cm (0.75 in). The wall thickness of the panels was nominally 3.18 mm (0.125 in). However, as is commonly the case in the blow-molding process, this wall thickness was not a well-controlled dimensional parameter. Since there is a mold surface on only the outside of each blow-molded wall, the local thicknesses of that wall can vary, depending on the geometry of the part. In this particular example, the typical variation in wall thickness is illustrated in Fig. 4.53 as measured from panels that were sectioned after fabrication.

It is common practice to measure mechanical properties such as Young's modulus for blow-moldable materials by conducting the test on an injection molded tensile specimen made from the plastic pellets. Use of this material data requires the assumption that the mechanical

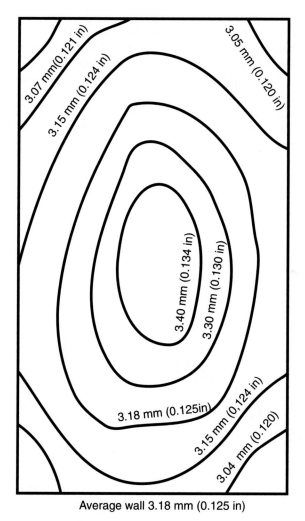

Average wall 3.18 mm (0.125 in)

Figure 4.53 Typical thickness variation of blow-molded panels.

properties of the injection-molded specimens are representative of the blow-molded material. In order to assess the accuracy of this assumption, stress–strain measurements based upon standard injection-molded tensile specimens as well as specimens cut directly out of the panel walls were compared. The specimens from the panel were cut in directions parallel and perpendicular to the extruding direction of the original parison in order to quantify any directionality introduced during this process. All the test material, both blow molded and injection molded, was from the same production lot.

The stress–strain behavior measured from both injection-molded and blow-molded specimens of M-PPE are compared in Fig. 4.54. The moduli measured from the specimens cut directly out of the blow-molded panels were slightly lower than the moduli from injection-molded bars. The bars oriented in the extrusion direction from the blow-molded specimens exhibited the lowest moduli. However, none of the differences, including those between the extrusion and cross-extrusion directions, are substantial, as can be seen from the figure. For this grade of M-PPE, there appears to be little difference between the injection-molded stress–strain data and data for the material cut directly from the blow-molded panel. Data reported based on the injection-molded bars should be sufficient to use for engineering stiffness predictions for this panel structure. However, it should be emphasized that

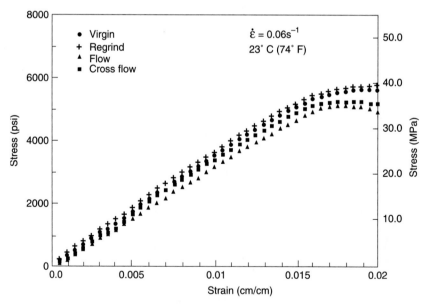

Figure 4.54 Comparison of stress–strain behavior as measured using injection-molded and blow-molded tensile specimens of a blow-molding grade of M-PPE.

the geometry of this application led to a minimum of stretching during the blow-molding process. Additional work similar to this investigation for situations where the part size requires larger amounts of stretching during molding is still required before a complete set of conclusions can be identified relative to the possible directional dependence of mechanical properties introduced during the blow-molding process.

In contrast to the data describing the material in the panel walls, accurate mechanical data for the rigid urethane foam is more difficult to generate and in general is much less consistent. The elastic properties such as Young's modulus and shear modulus are reported to be directly dependent upon the density of the foam, as shown in Fig. 4.55.[38] As can been seen from these figures, however, there is considerable scatter in the data. In order to best characterize the foam material used to fill the panels in this investigation, some of the panels were cut open and segments of foam were removed and subjected to compression tests. These tests indicated that the foam used here behaved in a bilinear fashion with an initial compressive modulus of 4.1 MPa (600 psi), an elastic limit of 0.1 MPa (15 psi), and a secondary modulus of 1.0 MPa (150 psi). With a modulus of 4.1 MPa (600 psi), this foam core is almost 3 orders of magnitude less stiff than the blow-molded thermoplastic. Similarly, the proportional limit is also nearly 3 orders of magnitude lower than that of the thermoplastic walls. As will be seen subsequently, these properties have a significant effect upon the performance of the panels.

Using the measured material properties for the blow-molded M-PPE and the urethane foam, which are summarized in Table 4.5, the following material modeling assumptions were employed to predict panel stiffness. A homogeneous, isotropic, linear-elastic material model was used to describe the M-PPE face sheets. Obviously, a multilinear representation of the stress–strain behavior could have been applied in a fashion similar to that employed in the box-beam example discussed previously. However, in this case, strains in the panel face sheets proved to be low enough that the difference was inconsequential. In the case of the urethane foam, a bilinear representation of the material stiffness was employed.

Having carried out fundamental material tests and established initial material models, a simple structural test was defined to assess the accuracy of stiffness predictions. For these tests, the 63.5×25.4 cm (25 $\times$ 10 in) panels were simply supported along all four of their sides and loaded in the center with a steel cylinder 5.1 cm (2 in) in diameter, as shown in Fig. 4.56. Both crosshead displacement and load were recorded. Since the parts tested were double-walled in nature, transverse displacement was measured under the load both at the top and bottom surfaces of the panel. This allowed local crushing beneath the load to be clearly identified. Crosshead displacement was used to quantify de-

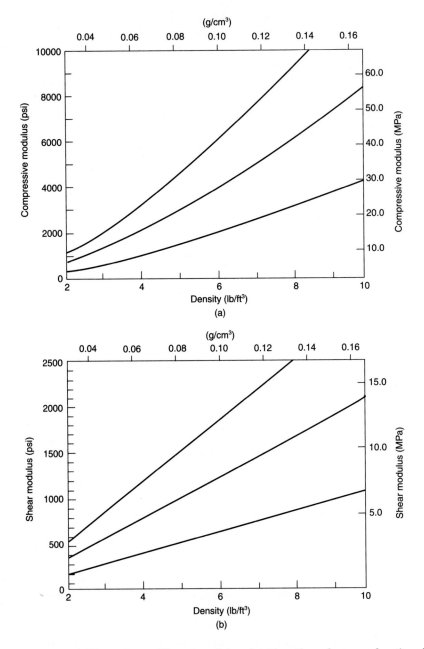

Figure 4.55 (*a*) Dependence of Young's modulus of rigid urethane foam as a function of density. (*b*) Dependence of shear modulus of rigid urethane foam as a function of density. Center lines represent average performance; top and bottom lines represent scatter limits.

TABLE 4.5   Summary of Mechanical Properties for Blow-Moldable M-PPE and Polyurethane Foam

Skin: Blow-moldable M-PPE	
Young's modulus	2.24 GPa (325,000 psi)
Yield stress	35.8 MPa (5,200 psi)
Core: Rigid urethane foam	
Young's modulus	4.1 MPa (600 psi)
Shear modulus	1.6 MPa (230 psi)
Compressive yield strength	0.1 MPa (15 psi)
Post-yield	$E_2 = 0.0$ (perfectly plastic)
Moduli considered	$E_2 = 1.03$ MPa (150 psi) (strain hardening)

flection at the top surface and a dial gauge was used to measure deflection at the bottom surface.

Because of the "solid" nature of a foam-filled panel, three-dimensional solid brick elements were used to predict its stiffness. The ANSYS® finite-element code[42] and its STIF45 element were utilized in this analysis. This element is an eight-node, three-dimensional, isoparametric element with three translational degrees of freedom at each node. A geometrically nonlinear analysis was employed in order to account for the effects of large rotations in the strain–displacement relations so often encountered in plastic structures. Using the previously discussed material models, including the bilinear strain hardening foam performance, the prediction of the panel's load–displacement behavior agrees quite well with experiment, as illustrated in Fig. 4.57. Observation of the panels after the removal of the load, as well as comparison of the measured displacements at the top and bottom panel surfaces, revealed that there was local crushing of the core of the panel

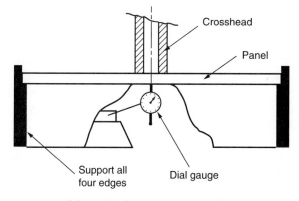

Figure 4.56   Schematic of test setup for panel testing.

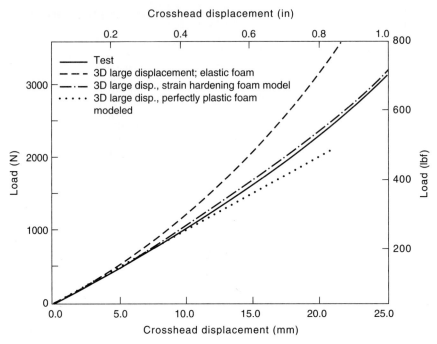

Figure 4.57 Comparison of predicted and measured stiffness of foam-filled panels.

under the load. Furthermore, the bilinear strain hardening material model for the foam core did play a role in the accuracy of the analysis. For comparison, predictions of the load–displacement behavior using a strictly linear material model for the foam as well as a bilinear, perfectly plastic model are also shown in Fig. 4.57. As can be seen, these predictions are less accurate. For the loads and displacements considered here, the thermoplastic face sheets never reached stress levels as high as the yield stress. As a consequence, a nonlinear description of the stress–strain behavior in the thermoplastic skins was unnecessary.

Having established the accuracy of the analysis, a number of parameter studies were carried out to determine which parameters most strongly control the panel stiffness. Of greatest significance in this work, summarized in Fig. 4.58, is the discovery that the overall panel stiffness for the low-density foam-filled panel can be significantly affected by increasing the distance between the panel face sheets. In Fig. 4.58, the panel stiffness, normalized by the stiffness of the baseline geometry, is reported as a function of three parameters—panel depth, skin thickness, and foam core modulus—all nondimensionalized by their respective baseline geometry values. The strong dependence of stiffness on panel depth is particularly important because it provides a

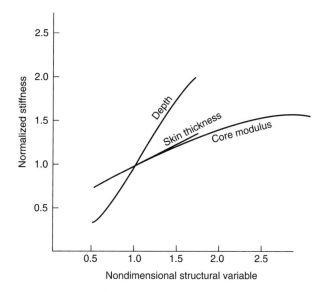

Figure 4.58 Normalized stiffness of foam-filled blow-molded panel as a function of nondimensional panel depth, skin thickness, and foam-core modulus.

method of increasing structural stiffness of a blow-molded panel when processing limitations make it impossible to increase wall thickness of a solid thermoplastic sheet. The panel stiffness is also significantly affected by increasing the modulus of the low-density foam core, in spite of the fact that it has a modulus orders of magnitude lower than that of the face sheets. This increase in the panel stiffness would appear to be related to improved shear transfer between the face sheets as the core stiffness is increased. As can be seen with reference to Fig. 4.55, the modulus of the low-density foam can be increased by increasing the density of the foam material. This in turn can be altered through foam process control.

Although the agreement between the numerical prediction using three-dimensional elements and the experimental measurements is quite good, there are practical reasons why alternative approaches to predicting the stiffness of such panels are desirable. The necessity of using several three-dimensional elements through the thickness adds complexity to the modeling process as well as increases the computer time required for analysis. Layered plate and shell elements offer an alternative for analysis of panels such as these. As an assessment of this approach for the low-density foam-filled core, layered plate elements, available in the ANSYS finite-element code (specifically the STIF53 and STIF91 elements), were used to predict the stiffness of the foam-filled panel. The analyses were carried out under the assumption of linear-elastic material behavior and small deflections. Consequently the results are accurate over a reduced range of displacements than those considered in Fig. 4.57. Two different finite elements were used in this assessment. The STIF53 element is a triangular layered plate element that ignores transverse shear deformation. The STIF91 element is an eight-noded, quadrilateral, layered plate element that includes transverse shear deformation. The panel was modeled using three layers—the two outside face elements and the core. The same linear elastic properties used with the three-dimensional elements (Table 4.5) were again employed. The results of these analyses are presented in Fig. 4.59 and compared to the results of the test and the three-dimensional analyses discussed previously. As can be seen, the laminated plate analyses are significantly stiffer (800%) than the three-dimensional analysis and the test results. The parameter study presented in Fig. 4.60 provides information that helps explain this discrepancy. In Fig. 4.60, the ordinate is the ratio of the panel stiffness as predicted using layered plate elements to the stiffness as predicted using three-dimensional elements. The abscissa is the ratio of core to skin moduli ranging from 10^{-3} to 1. As can be seen, for situations where the ratio of the core modulus to skin modulus is greater than 0.1, the results of the laminated plate and the three- dimensional analyses are much the same. However, as the core modulus becomes increasingly

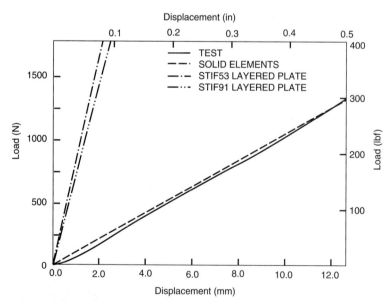

Figure 4.59 Comparison of predicted stiffness of foam-filled panel using three-dimensional finite elements and laminated shell elements with the ANSYS finite-element code.

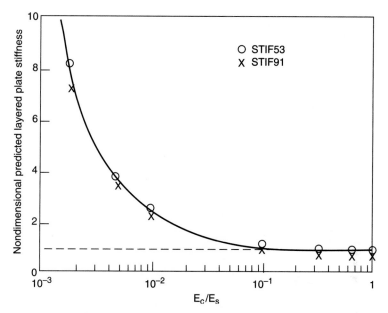

Figure 4.60 Layered plate model stiffness nondimensionalized by solid element model stiffness as a function of core-to-skin moduli ratio.

flexible in comparison to the skin modulus, the solutions deviate rapidly, with the laminated elements predicting much larger stiffnesses than the model using multiple three-dimensional elements through the thickness. Since the test results for the foam-filled panel correspond to a ratio of core modulus to skin modulus of nearly 10^{-3}, a significant discrepancy exists, and the three-dimensional solution has already been established to be more accurate based upon comparison with the test measurements. It is apparent that the assumptions regarding the distribution of strain through the thickness of the layered plate elements are not accurate for a panel with a very thick low-modulus core. In such cases, more specialized sandwich elements are necessary to predict the stiffness of such a panel adequately without resorting to detailed three-dimensional modeling through the thickness.

The second approach considered here to increase the stiffness of thermoplastic panels also utilizes a hollow, double-walled, blow-molded panel, but employs a different concept to couple the response of the two face sheets. It will be recalled that in this alternative approach, die inserts are pushed into one side of the panel until a portion of the wall on one side of the panel contacts the opposite wall. This process, referred to as a *tack-off process,* is carried out at several locations on the panel. In the case to be considered here, the die inserts are conical in nature. From a fabrication point of view, this approach to coupling the two face sheets of the hollow panel has advantages in both weight and cost in comparison to the foam-filled panels.

A detailed model of such a tack-off configuration is shown in Fig. 4.61. The finite-element model shown in these figures employs four-node quadrilateral shell elements and the ADINA finite-element code. Since a centrally applied load is considered, symmetry conditions have

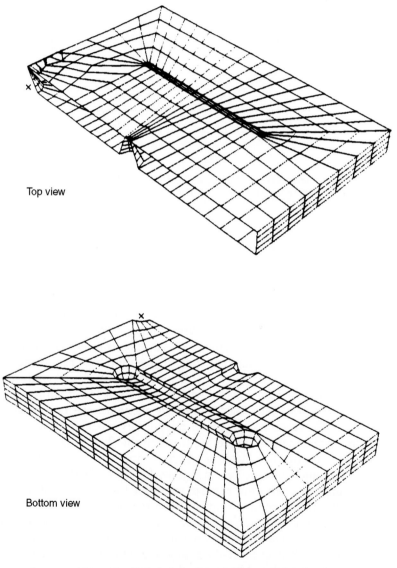

Top view

Bottom view

Figure 4.61 Finite-element model of blow-molded panels stiffened using the tack-off procedure (one-quarter of model shown).

been used to reduce the finite-element model to one-quarter of the panel. Figure 4.62 compares the results of a linear-elastic large-displacement analysis of this panel with load–displacement measurements from the simply supported plate configuration. For the range of behavior considered here, linear-elastic material properties for the thermoplastic were quite adequate since stresses were less than yield. The nonlinear nature of the curve is due to large rotation effects in the strain–displacement equations.

There are several points to be emphasized with respect to Fig. 4.62. First, the analysis compares very well with experiment, providing confidence in the accuracy of the modeling. The second striking observation is the significant reduction in stiffness apparent for this concept relative to the foam-filled panel in Fig. 4.57, in spite of the fact that the overall panel dimensions for this concept are the same as those for the foam-filled panel. The basic reason for this inferior stiffness performance of the tacked-off panel in comparison to the foam-filled panel is the inefficient shear transfer between the two face sheets. If the stress state in each face sheet of the tacked-off panel is examined, it can be seen that there is a gradient of stress from positive to negative in each face sheet. This indicates that each sheet is deforming as if it were an independent plate. If the tack-offs had been efficient in establishing

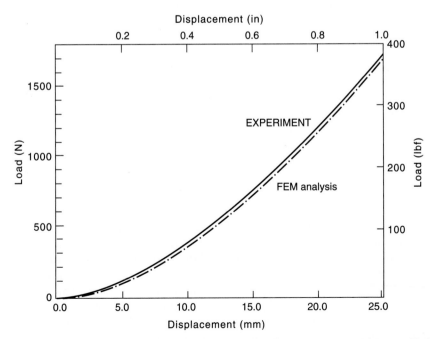

Figure 4.62 Comparison of predicted and measured stiffness of tacked-off blow-molded panels.

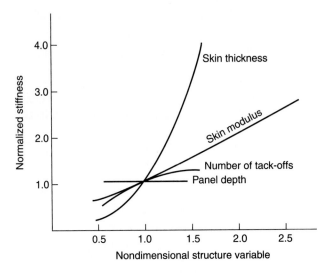

Figure 4.63 Normalized stiffness of blow-molded panel with tack-offs as a function of nondimensional skin thickness, skin modulus, panel depth, and number of tack-offs.

shear transfer between the two faces, then the upper face sheet would have been in compression throughout its thickness and the bottom face sheet would have been in tension.

This lack of efficiency in shear transfer is also visible in Fig. 4.63 where the numerical model is used to study the effect of various geometric alterations on the stiffness of the panel. The most notable result in this respect is that increasing the face-sheet separation does not improve the stiffness of the panel. Instead, the parameter that has the largest effect upon increasing the panel stiffness is increasing the skin thickness of each face sheet. More effective utilization of this "tack-off" fabrication technique may be possible. However, different configurations than those explored here would certainly be necessary.

Closure

The generally low material moduli associated with plastics makes design for stiffness an important issue for components made of polymers. For some homogeneous plastic materials, standard measurements of stress and strain are quite adequate for characterizing the material stiffness. These measurements can then be applied within the framework of finite-element analyses to predict stiffness of general part geometries accurately. In some situations, the nonlinear stress–strain behavior may need to be accounted for in order to reach high levels of accuracy. In many situations, geometric nonlinearity due to large rotations can also have a significant effect upon component stiffness.

On the other hand, there are numerous situations where less standard material measurements must be employed to characterize material stiffness fully. Material morphology and nonhomogeneity may lead to situations where standard tensile test data are inadequate in characterizing material stiffness for general engineering analysis. Structural foam is a good example of such a case. Because of the nonhomogeneous layers of this material, both bending and tensile tests using specimens cut from a foam-filled plate are necessary to establish mechanical data with general relevance. In situations where glass reinforcement is utilized, material properties may be strongly directional in nature requiring additional measurements to define orthotropic-elastic material behavior. In addition, these materials also require a means of defining material directionality due to processing effects. Design analysts must be alert to the potential of such situations when stiffness is a primary issue.

In spite of their low moduli, structural stiffness can be achieved in thermoplastic components by making use of some of the advantages of the many fabrication processes available. Thin-walled beams and panels can be manufactured with significant stiffness. However, it should be recognized that the large deformations of such thin-walled structures can lead to significant nonlinearity, which must be considered when designing and analyzing plastic parts.

References

1. *1992 Annual Book of ASTM Standards,* Vol. 08.01, *Plastics (1),* American Society for Testing of Materials, Philadelphia, PA, 1992, pp. 159–181.
2. I. S. Sokolnikoff, *Mathematical Theory of Elasticity,* McGraw-Hill, New York, 1956, pp. 56–65.
3. V. K. Stokes, "Vibration Welding of Thermoplastics—Part I: Phenomenology of the Welding Process," *Polymer Engineering and Science,* 28 (11): 718–727, 1988.
4. V. K. Stokes, "Vibration Welding of Thermoplastics—Part II: Analysis of the Welding Process," *Polymer Engineering and Science,* 28 (11): 728–739, 1988.
5. V. K. Stokes, "Vibration Welding of Thermoplastics—Part III: Strength of Polycarbonate Butt Welds," *Polymer Engineering and Science,* 28 (15): 989–997, 1988.
6. V. K. Stokes, "Vibration Welding of Thermoplastics—Part IV: Strengths of Poly(Butylene Terephthalate), Polyetherimide, and Modified Polyphenylene Oxide Butt Welds," *Polymer Engineering and Science,* 28 (15): 998–1008, 1988.
7. V. K. Stokes, "Analysis of the Friction (Spin)-Welding Process for Thermoplastics," *Journal of Materials Science,* 23 (8): 2772–2785, 1988.
8. V. K. Stokes, "Joining Methods for Plastics and Plastics Composites: An Overview," *Polymer Engineering and Science,* 29 (19): 1310–1324, 1989.
9. V. K. Stokes and S. Y. Hobbs, "Strength and Bonding Mechanisms in Vibration-Welded Polycarbonate to Polyetherimide Joints," *Polymer Engineering and Science,* 29 (23): 1667–1676, 1989.
10. V. K. Stokes, "Thickness Effects in the Vibration Welding of Polycarbonate," *Polymer Engineering and Science,* 29 (23): 1683–1688, 1989.

11. S. Y. Hobbs and V. K. Stokes, "Morphology and Strength of Polycarbonate to Poly(Butylene Terephthalate) Vibration-Welded Butt Joints," *Polymer Engineering and Science,* 31 (7): 502–510, 1991.
12. V. K. Stokes, "Strength of Glass-Filled Modified Polyphenylene Oxide Vibration Welded Butt Joints," *Polymer Engineering and Science,* 31 (7): 511–518, 1991.
13. R. P. Nimmer, O. A. Bailey, and T. W. Paro, "Analysis Techniques for the Design of Thermoplastic Bumpers," Society of Automotive Engineers (SAE) Technical Paper Series, Paper 870107, SAE, Warrendale, PA, 1987.
14. R. P. Nimmer, G. Tryson, and H. Moran, "Impact Response of a Polymeric Structure," *Proceedings of the 1984 Society of Plastics Engineers (SPE) Annual Technical Meeting,* SPE, Brookfield Center, CT, 1984, pp. 565–568.
15. V. K. Stokes and H. F. Nied, "Solid Phase Sheet Forming of Thermoplastics—Part I: Mechanical Behavior of Thermoplastics to Yield," *Journal of Engineering Materials and Technology,* Trans. ASME, 108 (2): 107–112, 1986.
16. J. L. Throne, "Structural Foams," in *Mechanics of Cellular Plastics,* N. C. Hilyard, Ed., MacMillan, New York, 1982, p. 263.
17. V. K. Stokes, "Design with Nonhomogeneous Materials—Part I: Pure Bending of Prismatic Bars," *Journal of Vibration, Acoustics, Stress and Reliability in Design,* Trans. ASME, 109 (1): 82–86, 1987.
18. V. K. Stokes, "Design with Nonhomogeneous Materials—Part II: Torsion of Thin-Walled Prismatic Bars," *Journal of Vibration, Acoustics, Stress and Reliability in Design,* Trans. ASME, 109 (1): 87–91, 1987.
19. V. K. Stokes, "Design with Nonhomogeneous Materials—Part III: Shear Effects in the Bending of Thin-Walled Prismatic Beams," *Journal of Vibration, Acoustics, Stress and Reliability in Design,* Trans. ASME, 109 (1): 92–96, 1987.
20. V. K. Stokes, R. P. Nimmer, and D. A. Ysseldyke, "Mechanical Properties of Rigid Thermoplastic Foams—Part I: Experimental Considerations," *Polymer Engineering and Science,* 28 (22): 1491–1500, 1988.
21. R. P. Nimmer, V. K. Stokes, and D. A. Ysseldyke, "Mechanical Properties of Rigid Thermoplastic Foams—Part II: Stiffness and Strength Data for Modified Polyphenylene Oxide Foams," *Polymer Engineering and Science,* 28 (22): 1501–1508, 1988.
22. D. A. Ysseldyke, R. P. Nimmer, and V. K. Stokes, "Computer-Aided Design and Analysis Procedures for Structural Foam Components," *Proceedings, SPI 14th Annual Structural Foam Conference,* Society of the Plastics Industry, NY, 1986; also *Plastics Design Forum,* HBJ, Denver, CO, July/August, 1986.
23. D. A. Ysseldyke, "Predicting the Structural Behavior of Thermoplastic Foam Parts," presented at the Sixteenth Annual SPI Structural Foam Conference, Society of the Plastics Industry, NY, April 1988.
24. Carl Zweben, H. Thomas Hahn, and R. Byron Pipes, *University of Delaware Composite Design Guide,* Vol. I—*Mechanical Behavior and Properties of Composite Materials,* Univ. Delaware, Newark, DE, 1985.
25. M. J. Folkes and S. Turner, "Alternative Testing Strategies for Short-Fiber Reinforced Thermoplastics," *Materials Design,* 6 (2), 1985.
26. Robert C. Wetherhold, William A. Dick, and R. Byron Pipes, *Thickness Effects on Material Properties in a Glass/Thermoplastic PET Injection Molding Compound,* SAE Technical Paper Series, Society of Automotive Engineers, Inc., Warrendale, PA, 1980.
27. P. S. Wright and A. Whelan, "Anisotropy in Short-Fiber Filled Thermoplastics," *Proceedings of the 1987 Society of Plastics Engineers (SPE) Annual Technical Meeting,* SPE, Brookfield Center, CT, 1987.
28. G. Ambur, "Material Characterization Methods and Structural Analysis Techniques for Injection Molded Glass Reinforced Thermoplastics," Masters of Engineering Thesis, Rensselaer Polytechnic Institute, Troy, NY, 1987.
29. G. Ambur and G. Trantina, "Structural Failure Prediction with Short-Fiber Filled, Injection Molded Thermoplastics," *Proceedings of the 1988 Society of Plastics Engineers (SPE) Annual Technical Meeting,* SPE, Brookfield, Center, CT, 1988, p. 1507.

30. G. D. Tomkinson-Walles, "Performance of Random Glass Mat Reinforced Thermoplastics," *Journal of Thermoplastic Composite Materials,* 1: 94, 1988.
31. V. K. Stokes, "Random Glass Mat Reinforced Thermoplastic Composites—Part I: Phenomenology of Tensile Modulus Variations," *Polymer Composites,* 11 (1): 32–44, 1990.
32. V. K. Stokes, "Random Glass Mat Reinforced Thermoplastic Composites—Part II: Analysis of Model Materials," *Polymer Composites,* 11 (1): 45–55, 1990.
33. V. K. Stokes, "Random Glass Mat Reinforced Thermoplastic Composites—Part III: Characterization of the Tensile Modulus," *Polymer Composites,* 11 (6): 342–353, 1990.
34. W. C. Bushko and V. K. Stokes, "Statistical Characterization of the Tensile Moduli of Random Glass Mat Reinforced Thermoplastic Composites," *Proceedings of the 1991 Society of Plastics Engineers (SPE) Annual Technical Meeting,* SPE, Brookfield, CT, 1991, pp. 2097–2101.
35. M. M. Matsco, "Nonlinear Finite Element Analysis in Plastics Design," *Proceedings of the 1989 Society of Plastics Engineers (SPE), Annual Technical Meeting,* SPE, Brookfield, CT, 1989, pp. 651–654.
36. K. C. Sherman, R. J. Bankert, and R. P. Nimmer, "Engineering Performance Parameter Studies for Thermoplastic Structural Panels," *Proceedings of the 1989 Society of Plastics Engineers (SPE) Annual Technical Meeting,* SPE, Brookfield, Center, CT, 1989, pp. 640–644.
37. *ADINA®* Theory and Modelling Guide, Report No. AE-84-4, ADINA Engineering, Inc., Watertown, MA, 1984.
38. D. A. Ysseldyke, "Investigation of Design Criteria and Analysis Techniques Applicable to Blow Molded Structures in Engineering Thermoplastics," Master of Science Thesis, Rensselaer Polytechnic Institute, Troy, NY, 1988.
39. D. A. Ysseldyke, "Influence of Processing on Mechanical Properties in the Design and Analysis of Blow-Molded Structures," *Proceedings of the 1989 Society of Plastics Engineers (SPE) Annual Technical Meeting,* SPE, Brookfield, CT, 1989, pp. 922–925.
40. D. A. Ysseldyke, "Design and Analysis of Low-Density Foam-Filled Blow Molded Panels," *Proceedings of the 1989 Society of Plastics Engineers (SPE) Annual Technical Meeting,* SPE, Brookfield Center, CT, 1989, pp. 966–971.
41. D. A. Ysseldyke, "Design and Analysis of Thermoplastic Blow Molded Panels with Tack-Offs," *Proceedings of the 1989 Society of the Plastics Industry (SPI) Structural Plastics Conference,* SPI, New York, 1989, pp. 83–87.
42. *ANSYS®* Engineering Analysis System Theoretical Manual, P. C. Kohnks, Ed., Swanson Analysis Systems, Houston, PA, 1989.

Chapter

5

Failure

In addition to providing adequate part stiffness, another significant concern in part design is usually preventing component failure. There are a number of ways that failure can be categorized. One approach to differentiating failure is on the basis of the loading condition under which failure occurs. Figure 5.1 is the simple illustration of the same four types of loading used earlier to aid in the outline of this text, and can also be used to categorize failures. In the following chapters (Chaps. 6, 7, and 8), failures associated with impact, creep, and fatigue will be discussed separately. However, for the introductory discussion of failure in this chapter we will deal with loads applied monotonically at a moderate rate until failure occurs. This type of loading is simply referred to as static maximum load in Fig. 5.1.

Component failure prediction is a much more difficult event to predict accurately than stiffness. There are several reasons for this increased level of difficulty. First of all, in order to predict failure, the deformation of the part must be accurately predicted. This means that attachments and other boundary conditions must be well defined as well as the load that is applied to the component. Although this may be possible in laboratory tests, it is usually much more difficult in realistic part configurations. In addition, it is clear from the preceding chapter treating stiffness that establishing an accurate capability of generally predicting the deformation of some polymeric materials is less than straightforward in nature. The discussions with respect to structural foam and glass-mat-reinforced composites are particular examples of some of the difficulties in establishing a general approach to measuring material properties that can be used to predict part stiffness accurately and generally. If there is uncertainty in prediction of the

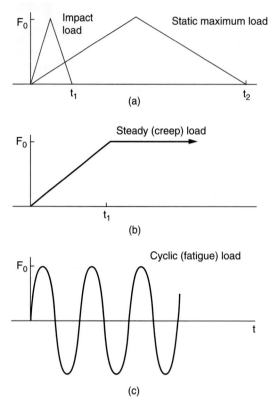

Figure 5.1 Different load conditions that can produce structural failure: (a) impact and static maximum loads; (b) long-term or creep load; (c) cyclic or fatigue load.

deformation process, it will naturally lead to less predictability in the failure process.

In general, failure of a component is defined by the performance requirements of the structure. Broadly speaking, failure can be considered to be the occurrence of any event defined to be unacceptable on the basis of the part's performance requirements. Within this context, failure might be considered to be excessive displacement under an applied load without the appearance of any damage at all. We actually dealt with the prediction of failure in this sense in the last chapter on stiffness. On the other hand, failure is often associated with some type of damage done to the material when the structure is subjected to excessive loads. In still other cases, damage may be tolerated but failure defined to be the inability of a structure to carry loads beyond some limiting value. Finally, in the worst case, failure can be a sudden and catastrophic event that leaves a structure in pieces. All of these last three categories of failure—damage, maximum load, and catastrophic breakage—will be discussed in this chapter.

Material choice and geometry definition are both important engineering decisions that will affect part failure. In order to predict failure in general part configurations, material properties must be available that define the failure limits. Usually, the only fundamentally useful material data available to assess expected failure limits for thermoplastics are simple tensile test data. This is not to indicate that these are the only required data; it is simply a statement of fact. In some circumstances, fracture toughness data may be available for some materials. When these data are correctly measured, they can be used to predict or assess failure of parts that already have a preexisting crack. However, practical experience in the effectiveness of using such data to predict failure of thermoplastic parts is very limited in the open literature at present, and not all published data on fracture toughness are equally reliable.

As part of the discussion of failure prediction in this chapter, several different types of damage mechanisms will be discussed. Some thermoplastics, especially those that are glass filled, display linear-elastic deformation until brittle failure at relatively low strain levels. This mode of failure will be the first subject of discussion in the chapter. Special attention is given to the issue of directionally dependent strength, which can be important for glass-filled injection-molded materials. On the other hand, many thermoplastics exhibit yielding and subsequent large plastic strain before failure. Although less catastrophic in nature, this type of damage accumulation, which is the second subject of discussion, can contribute to load limiting behavior as well as permanent deformation. However, even ductile thermoplastics with large strains to failure may undergo brittle catastrophic fracture in certain circumstances. One damage mechanism that must be mentioned in relation to brittle failure in thermoplastic materials is the appearance of crazes. A craze is quite similar to a crack in many respects except that in a craze, small fibrils of material connect what would normally be the free crack surfaces. The appearance of crazes has been observed to be related to brittle failure in many thermoplastics, and this will be the next subject of discussion in this chapter. An added complication to failure for many materials is the fact that both the quantitative value of the failure limit as well as the mode of failure for a particular material may change as a function of temperature. There are situations in which a material that is normally very ductile in nature may fail in very brittle and catastrophic fashion. Temperature dependence of failure will be discussed in this chapter. Cracks in plastic components can also develop through a variety of mechanisms, including fatigue, and can also lead to brittle failure. Experience in the engineering prediction of failure due to the presence of cracks will also be presented here. Finally, there are failure mechanisms that are not inherently material based. Structural instability and collapse is another mechanism that can act to limit the load-carrying ability of a structure and will be the final mode of failure discussed in this chapter. This mode of failure is somewhat unique in comparison to the other subjects because it is controlled by geometry and material modulus as opposed to limits in terms of ultimate strength or strain to failure. Each of these damage mechanisms as well as approaches to damage prediction will be discussed in this chapter and related to engineering failure of thermoplastic components.

Strength of Short-Fiber-Reinforced Thermoplastics

Glass-filled thermoplastics usually offer increased values of ultimate tensile strength as well as modulus in comparison to most unfilled thermoplastics. However, as was pointed out in the preceding chapter on stiffness, the standard test used to characterize this material employs injection-molded dog-bone specimens that induce fiber alignment along the specimen length. Strength measurements from these tests may not be representative of strengths in a component where preferred fiber alignment may not exist. The design engineer must be aware of the potential for reduced strength in other orientations with respect to flow and the decreased strain to failure of these reinforced materials. Although there does not appear to be extensive practical experience verifying any completely general approaches to predicting part strength for these materials, there are some practical approaches that attempt to address some of the recognized issues.

In the preceding chapter various studies of glass-reinforced thermoplastics were reviewed and an orthotropic elastic material model was presented and verified through tests. Similar test techniques are outlined here to establish material strength properties. These properties are then applied in simple geometries to predict part strength. Details of this approach are reported in Ref. 1.

To establish strength data that are more general with respect to flow orientation than the data from injection-molded bars, specimens were cut at various orientations from fan-gated plaques of different thicknesses. Table 5.1 provides a summary of the results for three glass-filled materials: 20% and 30% (by weight) glass-filled modified polyphenylene ether (M-PPE) and 30% glass-filled polybutylene terephthalate (PBT). The strength results in this table are normalized by the strength in the flow direction for the 3.2 mm (0.125 in) thick speci-

TABLE 5.1   Comparison of Normalized Ultimate Strengths Measured in Flow and Cross-Flow Directions for Three Glass-Filled Thermoplastics

[Ultimate Strength of Each Material Normalized by the Strength Measured in the Flow Direction from 3.2 mm (0.125 in) Thick Fan-Gated Plaques]

	Material				
	20% Glass-filled M-PPE	30% Glass-filled M-PPE			30% Glass-filled PBT
Specimen description	$t =$ 3.2 mm (0.125 in)	$t =$ 2.0 mm (0.08 in)	$t =$ 3.2 mm (0.125 in)	$t =$ 5.1 mm (0.2 in)	$t =$ 3.2 mm (0.125 in)
Flow direction (fan-gated plaque)	1.00	1.09	1.00	0.99	1.00
Cross-flow direction (fan-gated plaque)	0.78	0.58	0.61	0.65	0.55
Injection-molded ASTM specimen	1.21	...	1.03	...	1.29

mens for each material. For the 30% glass-filled materials, the cross-flow strength is about 60% that of the flow strength. In comparison, the end-gated injection-molded specimens often used to measure strength exhibit property values as much as 29% greater than strength of specimens cut in the flow direction of the plaque. Test results summarized in Fig. 5.2 for both 30% glass-filled M-PPE and PBT over a range of practical thicknesses show a trend of increased cross-flow strength relative to flow strength for increased thickness. However, for thicknesses of 5.1 mm (0.2 in) or less, the ratio of cross-flow to flow strength is 60% ± 10%.

The complete stress–strain curves for 30% glass-filled M-PPE in flow, cross-flow, and three intermediate orientations are shown in Fig. 5.3. For a brittle isotropic material, a maximum principal stress criterion can be applied to estimate component failure. However, for a material displaying process-induced directionality, a strength criterion is complicated by its dependence upon orientation of the material, as illustrated in Fig. 5.3. Since strength for an orthotropic material is of little use without definition of orientation, maximum stress as a function of orientation to flow direction at a specific point in a component must be determined in order to evaluate the strength of an orthotropic structure. This method is both complicated and time consuming.

Because of the complexity of using stress as a failure criterion in general components, an alternative method for evaluating the strength of an orthotropic material was developed. It was observed for the materials in this study that the end of the linear-elastic range (defined by the

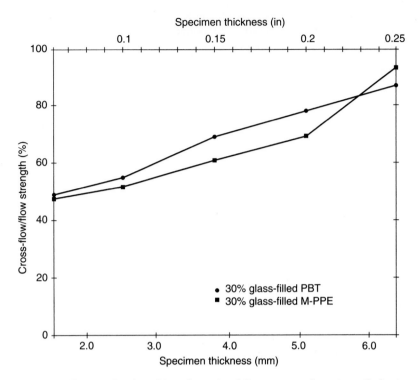

Figure 5.2 Test results describing the ratio of flow to cross-flow strength for two glass-filled thermoplastics as a function of plaque thickness.

proportional limit) for each orientation relative to the flow direction fell at approximately the same strain, as shown in Fig. 5.3. Hence, by using maximum principal strain instead of stress, the proportional strain limit at a point in a component is approximately independent of orientation.

Although use of traditional injection-molded dog-bone specimens to establish strength was shown to be very nonconservative, the proportional strain limit approach just outlined may be too conservative as a failure criterion for economic design, since there is substantial load-carrying ability remaining above the proportional limit. Therefore, as an empirical compromise, a practical strain design limit technique is suggested, allowing the engineer to extend the linear portion of the stress–strain curve beyond the proportional limit as shown in Fig. 5.4. It is suggested that tensile strength measured in the flow direction be divided by the elastic modulus associated with that direction to define a practical strain limit. Using this approach, the practical strain design limit is 1.4% for 30% glass-filled M-PPE (Fig. 5.3); 1.2% for 30% glass-filled PBT; and 1.6% for 20% glass-filled M-PPE. There will, of course,

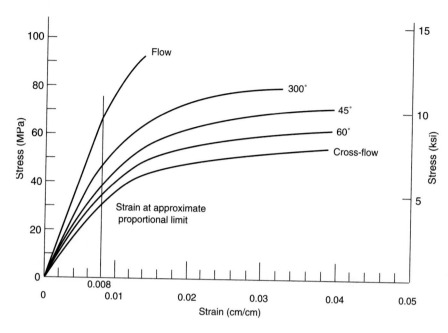

Figure 5.3 Stress–strain relationships measured for 30% glass-filled M-PPE as a function of orientation to injection flow.

be some error due to the nonlinear portions of the stress–strain curves that are ignored when using this approach.

In order to determine the accuracy of failure predictions obtained using the practical strain design technique, tests were carried out on flat plates loaded in bending and a cathode ray tube (CRT) housing. Measured strength results from these tests were then compared to strengths predicted from finite-element analyses. For comparison, analyses were also conducted that treated the material as isotropic and characterized by data from individually molded tensile specimens.

The flat-plate study consisted of cutting 10×15 cm (4×6 in) flat plates from injection-molded plaques in three orientations: flow, cross flow, and at a 54° angle to the flow direction. These plates were then tested in three-point bending until failure.

The CRT housing used for these tests had been specially designed so that it could be either edge or center gated. This part presented the opportunity to compare actual test results to structural finite-element analyses using two completely different flow patterns. Figure 5.5 shows the CRT housing and the two gate locations used. During the CRT-housing tests, all three vertical sides of the housing were constrained laterally and vertically by a fixture, and a point load was applied downward at the center of horizontal surface as shown in Fig. 5.5

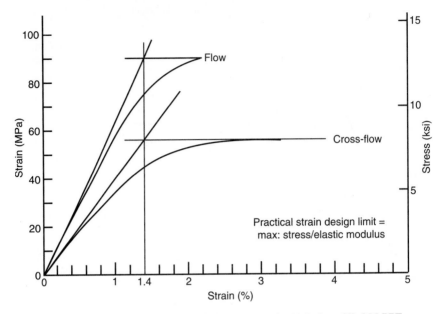

Figure 5.4 Practical strain design limit failure criteria for 30% glass-filled M-PPE.

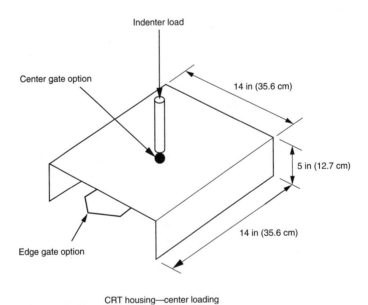

CRT housing—center loading

Figure 5.5 Test configurations for the CRT-housing strength tests.

until failure occurred. Because of its three-dimensional structure, the CRT housing is a more complex part than the flat plates. Additionally, the type of loading applied to the housing causes the part to experience large deflections, which give rise to geometric nonlinearities described in Chap. 3. Therefore, the complexity of both the geometry and loading of this part made it a more realistic test to assess the accuracy of the practical strain design limit and the orthotropic material model.

The same structural finite-element technique used in assessing the accuracy of stiffness predictions discussed in Chap. 4 was also applied here. This consisted of first determining flow direction vectors due to the injection-molding process for each element in the finite-element model of the component by performing a mold-filling analysis. The results were then used to define material directions for the orthotropic material model. A geometric nonlinear analysis was performed.

Table 5.2 reports the errors experienced in predicting measured fail-

TABLE 5.2   Comparison of Errors in Predicting Failure of a Beam Plate and a CRT Housing Using Isotropic Analysis with Injection-Molded Tensile Bar Strengths and Orthotropic Analysis with the Practical Strain Limit Criteria

Material, test condition	Load at failure	
	Isotropic prediction (% diff. from test)	Orthotropic prediction (% diff. from test)
CRT Housing		
20% glass M-PPE		
Edge gate	+34	−23
Center gate	+36	−28
30% glass M-PPE		
Edge gate	+56	−20
Center gate	+24	−17
30% glass PBT		
Edge gate	+155	−19
Center gate	+66	−35
Plate		
20% glass M-PPE		
Flow	+7	−14
Cross flow	+35	−14
54° angle	+3	−24
30% glass M-PPE		
Flow	−17	−19
Cross flow	+42	−21
54° angle	−3	−24
30% glass PBT		
Flow	−4	−32
Cross flow	+78	−21
54° angle	+39	−33

ure loads using the orthotropic analysis and practical strain limit as well as isotropic analysis and injection-molded bar data for both the flat beam plates and the CRT housing. Although the isotropic analyses were reasonably close in predicting failure loads of beams oriented in the flow direction, they were in error by as much as 78% in predicting cross-flow bending strength. The errors in using this approach to predicting the more complex and realistic CRT-housing failure were as high as 155%. The results tabulated in Table 5.2 emphasize how nonconservative this approach can be. On the other hand, predictions using the practical strain design technique yielded failure loads that

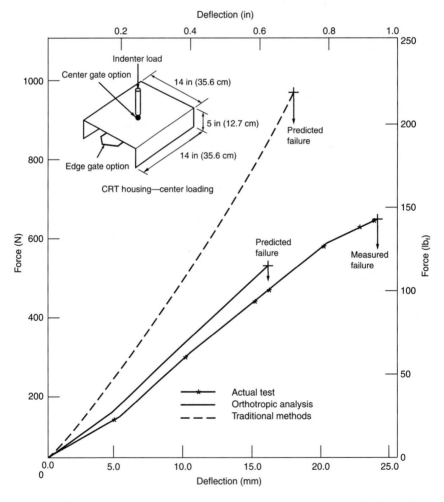

Figure 5.6 Comparison of predictions and measured test results for the load–displacement curve of the center-loaded, edge-gated CRT housing for 30% glass-filled M-PPE.

were consistently conservative by approximately 20%. Comparisons of predicted and measured load–displacement curves to failure of the CRT housing shown in Fig. 5.6 emphasize the relative accuracy of this approach. Of special additional interest is the fact that the orthotropic analysis coupled with a mold-filling analysis properly identified the strength advantage in the center-gate arrangement in comparison to edge gating the CRT housing for the 30% glass-filled M-PPE and PBT materials. The more standard isotropic analysis using injection-molded tensile bars would obviously predict the same strength for both arrangements, since it has no way of differentiating the effects of molding gate placement. Since this technique yields results that are both conservative and consistent, it appears to be a useful and practical tool for design engineers to predict failure in fiber-reinforced injection-molded thermoplastics.

Failure due to Yielding and Plasticity

Many plastic materials are capable of sustaining strains that are orders of magnitude larger than the few percent displayed by the glass-filled materials in the previous discussion. For materials such as polycarbonate, polypropylene, polyethylene, and polyetherimide, to name a few, the mechanisms of yield and plastic flow are very important relative to damage and failure.

The constitutive equations that form the basis of the theory of plasticity were originally developed to describe the behavior of metals, and thorough discussions of this behavior and the associated mathematical models used to describe it can be found in Refs. 2–4. Numerous careful experiments have been carried out that establish the validity of its application to describe the behavior of many metals. Although considerably less work of this nature has been performed for thermoplastics, it is generally accepted that the theory of plasticity is equally useful for many thermoplastics.[5]

There are two very important concepts associated with the theory of plasticity. The first concept is that of the yield limit. This limit, which is usually described in terms of stress, defines the theoretical boundary between the range of elastic behavior and plastic behavior. The second significant component of plasticity theory is the new relationship between stress and strain that governs the material's deformation behavior after yield. Generally, the plastic strain accumulated after yield is incompressible in nature and the incremental plastic strain components at any point in the material are proportioned the same as the vector components describing the normal to the yield surface at the associated point describing the current state of stress. Furthermore, the accumulated plastic strain is irreversible in nature and results in permanent deformation. This permanent deformation, of course, can be considered to be a form of damage. One of the important concepts behind the plastic constitutive model is that it defines how a damage parameter, plastic strain, is accumulated.

In most commonly applied plasticity models, the necessary material properties can usually be defined using a simple tensile test. The yield stress for metals is generally defined as the stress at which 0.2% irrecoverable plastic strain has been accumulated in a tensile test. Even though this yield-stress measurement is associated with a unidirectional stress field, there are mathematical generalizations available that allow the yield-stress limit to be defined for a multidirectional state of stress as well. One of the most popular is the von Mises yield criterion, which can be written as

$$\frac{1}{\sqrt{2}} [(\sigma_{xx} - \sigma_{yy})^2 + (\sigma_{yy} - \sigma_{zz})^2 + (\sigma_{zz} - \sigma_{xx})^2 + 6\tau_{xy}^2 + 6\tau_{yz}^2 + 6\tau_{zx}^2]^{1/2} = \sigma_y$$

$$(5.1)$$

where σ_{ij} and τ_{ij} represent the individual direct and shear stresses defining a general three-dimensional state of stress and σ_y is the yield stress measured in a standard unidirectional tensile test. The left-hand side of Eq. (5.1) is commonly referred to as the von Mises or effective stress.

Using this relationship, yield is defined to occur whenever the effective stress at a point in a component reaches the value of the unidirectional yield stress as defined by Eq. (5.1). Yee and Carapellucci[6] have investigated this condition for polycarbonate and found it to be approximately correct, although they identified noticeable orthotropy in the yield behavior of some specimen geometries, as is often observed in metals. Plasticity theories that account for this orthotropy are available but not often contained in commercial finite-element codes. In addition, very little experimental data are available to characterize this behavior for thermoplastics. As a result, isotropic yield criteria are most often applied. Figure 5.7 compares the two-dimensional version of Eq. (5.1) with experimental data from Ref. 6. Obviously, the availability of a generalized yield condition such as Eq. (5.1) is a very powerful tool in analyzing real components. Although few other comparisons similar to Ref. 6 exist, the von Mises yield criterion is often applied for many thermoplastics.[5]

It has been recognized[5] that the yield stress of some thermoplastics is dependent upon the hydrostatic stress (first invariant of the stress tensor). Some of the commercially available analysis codes offer the capability to account for this dependence. However, very little experimental data quantifying this dependence for engineering thermoplastics

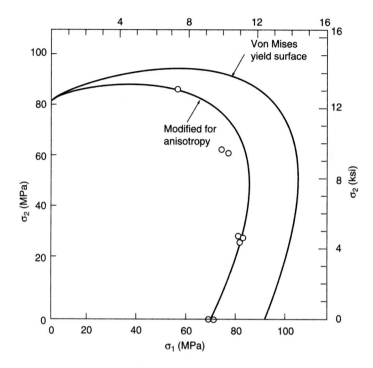

Figure 5.7 Comparison of measured biaxial yield states with the von Mises yield condition (Ref. 6).

exists. Therefore, in most cases, analysts will have to be satisfied to use more standard pressure-independent yield theories.

Once the yield limit is reached, the linear-elastic modulus is no longer sufficient in relating stress to strain. In some materials, although the new local slope of the stress–strain curve is significantly less after yield, the relationship is still approximately a linear one. In such cases, a bilinear relationship between stress and strain, as shown in Fig. 5.8, may be appropriate. In other cases, the post-yield stress–strain behavior may be very nonlinear and a multilinear approximation (also shown in Fig. 5.8) may be more appropriate. Both of these models are available in many of the standard finite-element codes.

Observed tensile test behavior in ductile thermoplastics

The relationship between stress and strain after yield is also usually defined for most engineering materials using unidirectional tensile test data. These data can generally be measured using a load cell and standard extensometer as long as the deformation is homogeneous

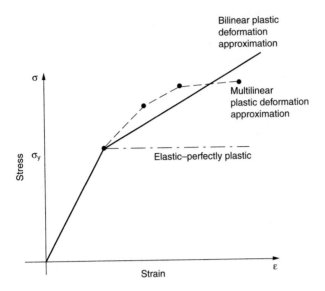

Figure 5.8 Bilinear and multilinear approximations to elasto-plastic constitutive behavior.

over the specimen gauge section. For many materials, a state is eventually encountered during plastic deformation when a localization of the strain field known as *necking* occurs. Such an event is illustrated schematically in Fig. 5.9. After such an event initiates, normal extensometer data are generally useless since the location of the extensometer with respect to the localization becomes a significant unknown. In addition, the calculation of the stress in the material must generally be made to reflect the change in cross-sectional area of the specimen. It is inadequate to simply use the original cross-sectional area during localization events.

Although yield limits and classical plasticity have been found to be very useful in engineering calculations for thermoplastics as well as metals, there are some important differences to be noted in the yield behavior of plastics in comparison to other ductile materials. Figure 5.10 illustrates the true-stress versus stretch behavior as measured for three thermoplastics—polycarbonate (PC), polyetherimide (PEI), and polybutylene terephthalate (PBT)—in a tensile test and reported in Ref. 7. The stretch parameter λ in this figure is defined as

$$\lambda = \frac{d\overline{s}}{ds} \tag{5.2}$$

where ds is the length of a material segment before deformation and $d\overline{s}$ is the length of the same segment after deformation. As visible in

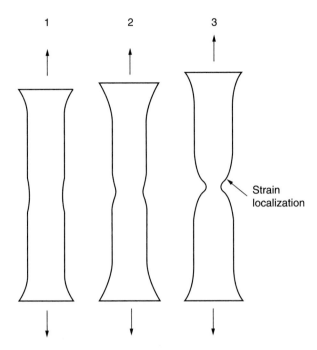

Figure 5.9 Schematic illustration of a strain localization process in an axisymmetric tensile specimen.

Fig. 5.10, polycarbonate can be stretched to lengths on the order of 200% of its original dimensions before breaking in a tensile test. Polybutylene terephthalate reaches stretch values of almost 350%. However, these large stretch values are achieved in a very nonlinear manner and there are several events that occur well before the final rupture that might constitute engineering failure in the more general sense.

The first of these events is the initial maximum that occurs in the stress-versus-stretch curve. For small values of stretch, λ can be related to engineering strain as

$$\varepsilon = \lambda - 1 \tag{5.3}$$

Using this relationship, the initial portions of the polycarbonate stress–stretch curve can be recast in the more recognizable engineering form of the stress–strain curve shown in Fig. 5.11.

There are several characteristics worth discussing relative to this small strain range of polycarbonate's behavior. First, it can be noted that the stress–strain relationship is not linear all the way to the maximum in the curve. Nonlinear behavior initiates at stress levels between 50% and 60% of the initial maximum stress. During the

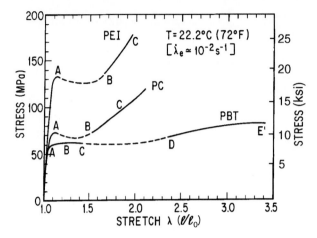

Figure 5.10 True-stress versus stretch relationship for polycarbonate, polyetherimide, and polybutylene terephthalate.

investigation by Stokes and Nied,[7] which led to the stress-versus-stretch data reported in Fig. 5.10, it was noted that in spite of the non-linearity in the early regions of the stress–strain curve, there was no permanent deformation accumulated in the tensile specimen for stress levels less than the maximum value. Whenever the specimen was unloaded at lower loads, all of the strain was recovered, although unloading from load levels near the maximum led to time-dependent recovery of some of the strain. Consequently, on the basis of permanent defor-

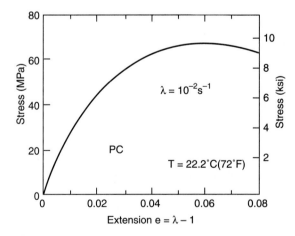

Figure 5.11 Initial region of the polycarbonate stress–strain curve.

mation onset, Stokes and Nied[7] suggest use of the initial maximum in the stress–strain (or stress–stretch) curve to define the yield stress and onset of permanent deformation for the materials they studied.

The initial maximum (yield) stress in engineering thermoplastics is very dependent upon temperature. In general, the yield stress decreases with increasing temperature as shown in Fig. 5.12, which represents data from Ref. 7 describing yield stress as a function of temperature for polycarbonate, polyetherimide, and polybutylene terephthalate.

After the initial maximum in the stress–stretch curve shown in Fig. 5.10, many thermoplastics such as PC, PEI, and PBT exhibit quite different behavior than common ductile metals. Beginning with local shear banding near the specimen edges, the cross section of the material rapidly begins to thin and exhibit strain localization, apparently quite similar to the necking process that occurs in metals. However, this initial necking is followed by a cold-drawing (neck propagation) process in the thermoplastics, which differentiates its behavior from common structural metals. Robertson[8] suggests that the necking and cold-drawing of plastics are often characterized by two features: the shape of the load–extension curve in tension and the shape of the drawn test specimen. Both of these characteristics are exhibited for polycarbonate in Fig. 5.13. The most visually striking aspect of this be-

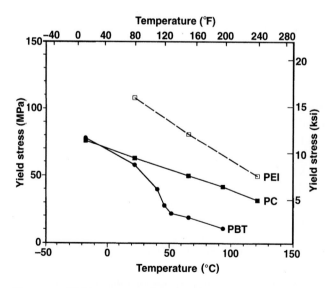

Figure 5.12 Yield stress as a function of temperature for polycarbonate (PC), polyetherimide (PEI), and polybutylene terephthalate (PBT) tested at a strain rate of 10^{-3} s^{-1}.

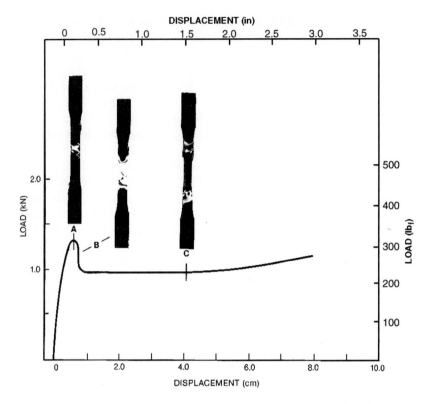

Figure 5.13 Load–displacement behavior associated with necking and neck propagation in polycarbonate.

havior is the fact that the dimensional thinning process associated with necking in some thermoplastics does not remain local as it does in metals. Instead, thinning at the original necked area eventually terminates and the local deformation process there stabilizes. Once localization stabilizes at the original neck location, thinning initiates at the boundaries of the necked area, and the necked area begins to grow toward the tensile specimen grips as shown in the sequence of photographs in Fig. 5.13 from a test carried out on polycarbonate. This process is referred to as cold-drawing or neck propagation. Most of the early material science investigations directed at cold-drawing in polymers, for example,[8–11] dealt primarily with the thermodynamic and molecular mechanisms that may initiate the yield process at point A in Fig. 5.13, rather than the mechanics of a stable, propagating neck. Vincent[12] appears to have been one of the first to discuss the mechanical aspects of propagating necks in polymers. He argues that the drop in load after yield at point A in Fig. 5.13 is related to the mechanical process of cross-sectional area reduction at the neck. Furthermore, the construction originally applied by Considere is used to argue that a neck of finite cross-sectional area will propagate along the length of a specimen only if the local hardening modulus at large strain exceeds the true stress at that strain. Regarding the decrease in load after yield, Brown and Ward[11] argue that the reduction observed in polymers cannot be explained purely on the basis of the mechanics of necking. They emphasize that during tensile tests of polyethylene terephthalate, a decrease in true stress was observed after yield. Other investigators have also subsequently reported similar decreases in true stress in polycarbonate[7,13–16] and polyvinyl chloride[17] and high-density polyethylene[17] and polyetherimide.[7,13]

Initial yield and maximum load

Having discussed the sequence of events generally observed in the tensile test of many common ductile thermoplastics, the relationship of these events to engineering failure can be examined. As an example of predicting failure associated with initial yielding in components made of thermoplastic, we will use the simple two-dimensional box section[18] introduced in the last chapter and illustrated in Fig. 5.14. It will be recalled that the material used to manufacture this box section was a rubber-toughened polycarbonate. The stress–strain relationship for this material as measured in simple tensile tests is shown in Fig. 5.15 for a strain rate of 0.1 s^{-1}. As mentioned earlier, when an extensometer is used to measure strain, the data are only valid for loads prior to localization. As a result, only data up to the maximum stress point of the curve are presented in this figure. The same finite-element model used to predict the stiffness of this part (illustrated in Fig. 4.11) was applied to predict damage and failure due to yielding. Since the box section shown in the figure was 10 cm (4 in) in length perpendicular to the plane of Fig. 5.14, plane-strain analysis was carried out.

   The material model applied in this analysis was also previously discussed in Chap. 4. It is elastoplastic in nature and represents the nonlinear relationship between stress and strain with a piecewise linear approximation, as shown in Fig. 5.15. It was shown in Chap. 4 that this material model improved the accuracy of the load–displacement prediction at higher loads. There are several approximations associated with using this material model to represent the behavior of polycarbonate. First, when using this model, the yield point defining the onset of irrecoverable plastic deformation is associated with the proportional limit of the stress–strain behavior. Although this approximation is inconsistent with the observation of permanent deformation beginning at

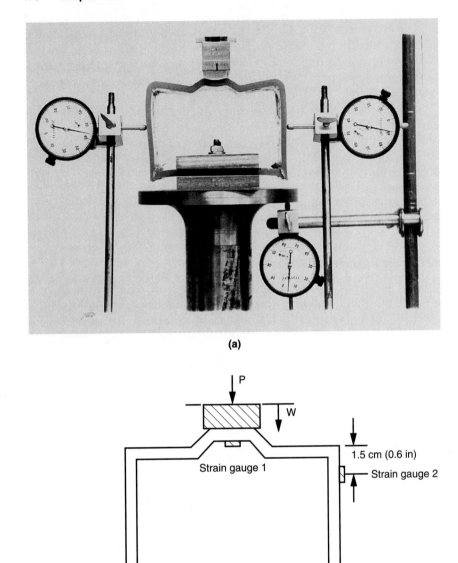

(a)

(b)

Figure 5.14 Photograph (a) and diagram (b) of setup for two-dimensional box-section test.

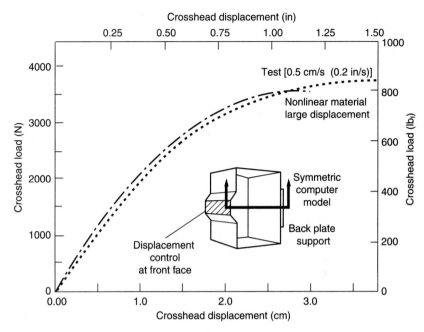

Figure 5.20 Comparison of experimentally measured and analytically predicted maximum loads for the box test.

Chap. 4 to encompass the full range of response to maximum load. The most important point to emphasize in this comparison is the accuracy of the prediction for maximum load in this test (<10% error). The appearance of such a maximum load is clearly a possible failure criterion for a structure where the load is controlled. Since stress levels cannot exceed the maximum value of about 57 MPa (8.3 ksi) before reaching a local maximum in the stress–strain curve shown in Fig. 5.15, a maximum load for the structure is eventually reached as deformation increases. Furthermore, the approximation of perfect plasticity after the maximum stress level is reached in the stress–strain curve does not appear to have introduced any substantial errors within the context of engineering calculations. As can be seen in Fig. 5.17, the strain at the point of highest stress and deformation does not significantly exceed 6% at the maximum load level measured in the component.

For materials such as polycarbonate, this approach of using classical elastoplastic theory is obviously quite useful in predicting failure associated with permanent deformation or maximum load in an engineering component. However, some care must be taken in applying the constitutive relationships of plasticity to any thermoplastic that displays a stress–strain curve that is nonlinear in nature. One of the basic assumptions of classical plasticity is that the deformation process is incompressible. Many thermoplastics display what appears to be a yield stress followed by strain accumulation without change in measured stress level. However, in some cases the deformation process after maximum stress is reached is clearly not incompressible. Internal voiding may take place as deformation continues beyond the yield point. In such a case, the relationship between actual behavior and predictions made with plasticity theory remain less clear.

Strain localization, necking, and neck propagation

Although the yielding behavior of polycarbonate led to a maximum load at strains near yield value in the box-section component considered in the last section, such behavior is not necessarily guaranteed for all geometries. There are some situations where maximum load and failure of plastic are associated with the materials' behavior well after the yield strain is surpassed. Although this is most often a very severe loading situation, one example of such behavior does have an important relationship to a test often used to characterize the strength of plastics—the puncture test illustrated schematically in Fig. 5.21. As a result, this type of failure merits discussion in spite of the fact that the modeling necessary to predict such behavior does not fall within the context of normal engineering calculation.

In order to understand and predict failure in general component geometries made of ductile thermoplastics at these very large strains, it must be possible to model the physical events of necking, drawing, and

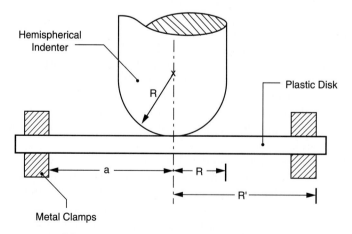

Figure 5.21   Schematic of a thermoplastic puncture test.

neck stabilization, which are observed in the tensile specimen. For many thermoplastics, these events begin, from a material point of view, at the maximum point in the true-stress versus stretch curves shown in Fig. 5.10. For thermoplastics such as polycarbonate, this is not only the point at which irreversible plastic deformation begins, but also localization and drawing. Some other thermoplastics like PBT behave somewhat differently. In the case of PBT, a semicrystalline polymer, there is a range of strain following yield during which homogeneous plastic deformation occurs. Strain localization begins at much higher strains on the order of 20%.

The processes of necking and subsequent stable neck propagation that are observed in thermoplastics such as polycarbonate, polyetherimide, and high-density polyethylene have been discussed by Ward.[20] There are also a number of detailed observations reported in the literature with respect to specific thermoplastics.[7,13,14] In addition, a number of people have investigated general constitutive equations relating stress to strain with the object of identifying the fundamental relationships necessary to model the characteristic behavior of neck propagation that is observed in many plastics. Such equations are necessary in order to extend the information available from tensile test data to more general geometries. The strain-hardening behavior of thermoplastics like polycarbonate, which begins at a stretch of about 1.4 in Fig. 5.10, is a very significant contributor to the phenomenon of stable neck propagation. Hutchinson and Neale[21] apply standard plasticity equations to study this propagating neck phenomenology and illustrate analytically that strain hardening at large strain values can lead to a propagating neck as seen in plastics. Nimmer and Miller[22] also make use of standard plasticity relationships to carry out parameter studies of several generic material properties defining the true-stress versus true-strain dependence of thermoplastics and their effect upon observed phenomenology in the tensile testing of plastics. Coleman and Newman[23] have shown that very similar propagating neck behavior is predicted if constitutive equations describing incompressible rubber elasticity are applied instead of irreversible plasticity. Based upon physical concepts related to the amorphous macromolecular structure of polymers and the microstructural mechanisms by which plastic deformation is accommodated in these materials, a rate-independent constitutive model for large elastic-plastic deformation of glassy polymers was developed by Bagepalli, Argon, and Parks.[24-26] Within the framework of this model, the molecular orientation associated with large plastic stretch eventually leads to the development of a rubberlike internal resistance to deformation. Boyce, Parks, and Argon[27] have extended this model to treat additional phenomena important to plastics including rate sensitivity. Batterman and Bassani[28] have attempted to include the effects of pressure sensitivity and strain-induced anisotropy in their constitutive models. These models account for many of the phenomena observed in thermoplastics and are formulated for general application in finite element codes, although they are not available in the more common commercial codes.

While the details of the modeling used in Refs. 21 to 28 to treat the phenomena of propagating necks in plastics are relatively complex in comparison to normal design analyses, a general discussion of some of the underlying physical behavior is useful. Such understanding may also provide a vehicle to generalize conceptually some of the results to other problems where similar behavior is an important issue.

In order to examine the relationship between basic material properties and the observed behavior in the tensile tests of some ductile thermoplastics, an axisymmetric, tensile specimen with applied displacement loading shown in Fig. 5.22 is modeled using eight-node, two-dimensional isoparametric finite elements and the ADINA® nonlinear finite-element code that includes the effects of finite deformation. The

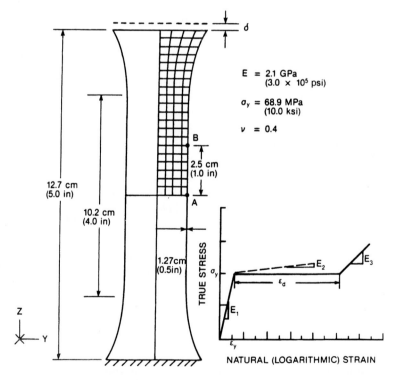

Figure 5.22 Finite-element model of an axisymmetric tensile specimen under applied displacement loading.

details of this analysis can be found in Ref. 22. Figure 5.22 also illustrates the finite-element mesh used for the numerical results presented here. As can be seen in this figure, the shoulder of the tensile specimen was included in the model. As a by-product of this geometry, necking in the analysis always initiated at the center of the specimen without the necessity of including any geometric or material imperfection. In reality, the necking process may initiate anywhere in the specimen. Except where explicitly stated in the text, the material properties used for these analyses are also listed in Fig. 5.22 and are representative of a ductile polymer such as polycarbonate.

The mechanical relationships between constitutive behavior, onset of necking, and post-yield stability of the axisymmetric tensile specimen can be studied with the simple bilinear, elastoplastic constitutive model also shown in Fig. 5.22. Figure 5.23 illustrates variations in the predicted load versus crosshead displacement behavior of the tensile specimen as a function of the initial hardening modulus E_2 in the bilinear constitutive model. With elastic modulus and yield stress held constant at the values listed in Fig. 5.22, there is a significant difference in the post-yield, load–displacement behavior as the value of E_2 is varied from $+2\sigma_y$ to $-2\sigma_y$. For a hardening modulus of 140 MPa (20 ksi),

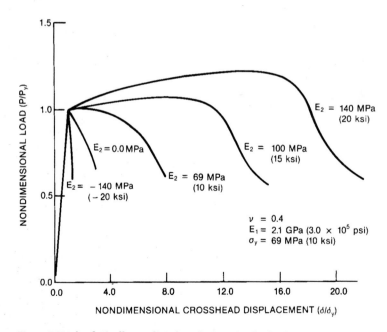

Figure 5.23 Analytically predicted variations in the load versus crosshead displacement behavior of a tensile specimen as a function of the initial hardening modulus E_2.

a value twice the yield stress, the load–displacement behavior in Fig. 5.23 is similar to that seen in many metals (titanium, for example). As the load increases after yield, the deformation in the tensile specimen is homogeneous and plastic in nature. Based on the principle of incompressible plastic strain, it is well known that a maximum in the load–displacement curve is expected when the true stress in the specimen reaches a value equal to the hardening modulus $d\sigma_t/d\varepsilon_t$. At that point, a neck initiates and the load begins to decrease. When the hardening modulus E_2 is reduced to 69 MPa (10 ksi), a value equal to the material's yield stress, both the load–displacement and the deformation behaviors of the tensile test are characteristically different. The maximum in the load–displacement curve now takes place at the yield point and the load decreases as the crosshead displacement increases in Fig. 5.23. In addition, there is no longer any region of homogeneous plastic deformation. Instead, localized necking occurs simultaneously with yield. As the hardening modulus is reduced below the value of the yield stress, the maximum load and neck formation are always coincident with initial yield, and the post-yield slope of the load–displacement curve becomes increasingly negative. This type of behavior is often visible in the load–displacement behavior of ductile polymers.

For all values of E_2 considered in Fig. 5.23, the load decreases monotonically after initiation of a neck, the necked region remains local, and the cross-sectional area of the neck decreases continuously. As previously mentioned, a true-stress versus true-strain relationship characterized by increasing values of the post-yield modulus as a function of true strain is necessary for a neck to stabilize and propagate. To quantify the effects of this mechanism, a trilinear relationship between the true stress and true strain shown in Fig. 5.22 is applied to the axisymmetric tensile test. For this investigation, values for Young's modulus ($E_1 = 2.1$ GPa; 300 ksi) and yield stress ($\sigma_y = 69$ MPa; 10 ksi) characteristic of polycarbonate were chosen and the secondary modulus was fixed at $E_2 = 0.0$. The third-stage modulus E_3 and the amount of true strain between yield and third-stage hardening, ε_d, were varied parametrically.

Figure 5.24 illustrates the predicted load–displacement behavior in the axisymmetric tensile specimen illustrated in Fig. 5.22 for $\varepsilon_d = 0.4$ and a range of third-stage moduli from $E_3 = 0.0$ to $E_3 = 410$ MPa (60 ksi). The case of $E_3 = 0.0$ corresponds to the linear-elastic, perfectly plastic, bilinear case discussed earlier in Fig. 5.23 and shows a monotonically decreasing load associated with local unstable necking after yield. However, as E_3 is increased to values of 140 MPa (20 ksi) and greater, the load reaches a finite minimum and then remains constant for the crosshead displacements plotted in Fig. 5.24. Figure 5.25 illustrates the effect of the third-stage hardening modulus (E_3) in limiting

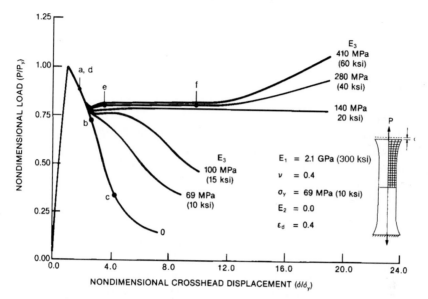

Figure 5.24 Predicted nondimensional load displacement behavior in an axisymmetric tensile specimen as a function of the final-stage hardening modulus E_3 in the trilinear stress–strain model.

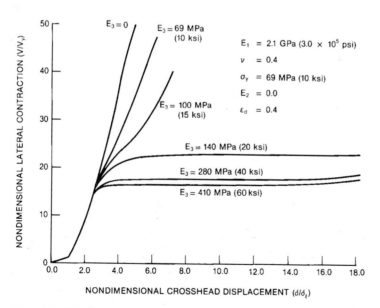

Figure 5.25 Nondimensional lateral contraction at point A in Fig. 5.22 as a function of the final-stage hardening modulus E_3 in the trilinear stress–strain model.

the decrease in cross-sectional area at the neck. Here, nondimensional lateral contraction at the location of neck initiation (point A, Fig. 5.22) is shown as a function of the nondimensional crosshead displacement. For values of E_3 less than or equal to 100 MPa (15 ksi), the lateral contraction at point A increases without bound. However, for values of E_3 greater than or equal to 140 MPa (20 ksi), there is a limiting value for the lateral contraction of the neck. For values of E_3 greater than 280 MPa (40 ksi), neither the load–displacement curve in Fig. 5.24 nor the lateral contraction versus displacement curve in Fig. 5.25 shows much additional variation as a function of E_3 until the neck has propagated the entire gauge length and the load begins to increase. Figure 5.26a to c illustrates the deformation of the specimen as predicted with the trilinear curve in Fig. 5.22 at three points in the load history of a material with $E_3 = 280$ MPa (40 ksi). The crosshead displacements and loads corresponding to these points are defined in Fig. 5.24 by the letters d, e, and f, respectively. In Fig. 5.26a, the specimen has yielded, the neck is just beginning to form, and the load is decreasing. The necking process continues with local decrease in the neck cross-sectional area until true strains larger than $\varepsilon_y + \varepsilon_d$ are incurred and third-stage hardening begins in the original neck area as illustrated in Fig. 5.26b. The increased stiffness of the material in the original neck cross section stabi-

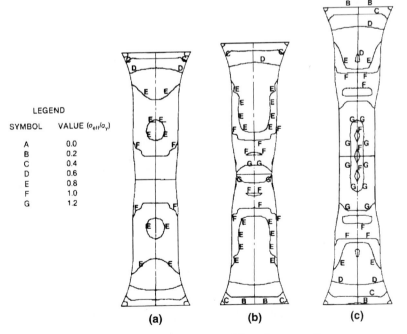

LEGEND

SYMBOL	VALUE (σ_{eff}/σ_y)
A	0.0
B	0.2
C	0.4
D	0.6
E	0.8
F	1.0
G	1.2

(a) (b) (c)

Figure 5.26 Effective stress contours during stable neck propagation.

lizes the necking process there and forces adjacent material to yield. As this process continues, the neck propagates along the length of the specimen as shown in Fig. 5.26c. When the neck covers the entire gauge section of the specimen, the final increase in load visible in Fig. 5.24 will occur prior to failure.

The draw strain, defined as ε_d in Fig. 5.22, also plays a role in the post-yield deformation and load-carrying capability. Figures 5.27 and 5.28 illustrate some of these effects. The results displayed in these figures are based upon a trilinear material model defined by $E_1 = 2.1$ GPa (300 ksi), $\sigma_y = 69$ MPa (10 ksi), $E_2 = 0$, and $E_3 = 280$ MPa (40 ksi). The draw strain ε_d is varied between 0.00 and 0.60. Lower values of ε_d result in smaller load reductions prior to stabilization and less overall crosshead displacement prior to final stiffening, as can be seen in Fig. 5.27. The smaller reductions in crosshead load before stabilization are, of course, associated with less cross-sectional area reduction in the neck, as can be seen from the curves of nondimensional lateral contraction versus nondimensional displacement in Fig. 5.28. For the case of $\varepsilon_d = 0$, the trilinear model degenerates to a bilinear relationship. In the case considered here, since the secondary modulus after yield is now greater than the yield stress, there is no load drop or necking phenomenon. Instead, homogeneous plasticity occurs throughout the specimen as discussed in relation to Fig. 5.23.

The presence or absence of this stable neck propagation behavior can

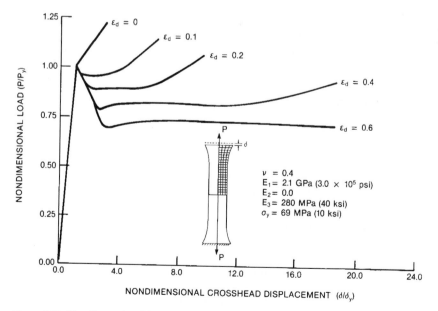

Figure 5.27 Nondimensional load–displacement behavior as a function of draw strain ε_d.

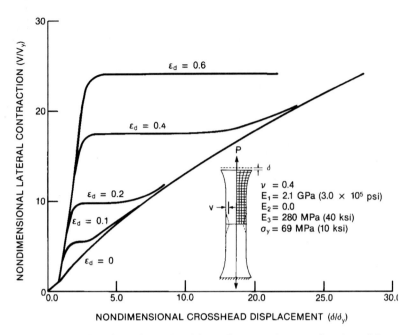

Figure 5.28 Predicted nondimensional lateral contraction as a function of draw strain ε_d.

have significant effects upon the failure of plastics. Yielding was previously discussed as one measure of failure that can be used in considering performance of thermoplastic components. It is associated with permanent deformation of the material and it can cause load-limiting behavior as was seen in the previous example of the thermoplastic box section. On the other hand, the onset of yielding certainly is not always synonymous with ultimate failure. Although higher loads may be prohibited, substantial additional deformation may occur without part rupture, as can be seen from the simple tensile test examples presented here.

In metals, the onset of necking is often considered a failure criteria. For those materials, the common situation is an initial post-yield modulus that is larger than the yield stress. As a result, as crosshead displacement increases, homogeneous plasticity prevails after yield as the deformation mode in the tensile specimen. When the stress level reaches values equal to the post-yield modulus, localization (necking) occurs. Since metals do not generally exhibit post-yield moduli that increase as a function of strain after yield, localization generally leads to final failure. If the strain state is two- or three-dimensional, this may occur at different strain values, and maps defining the onset of localization as a function of strain state have been used to define failure in metal-forming operations[29] as well as severe mechanical loading events such as bird-strike damage on fan blades made of ductile metals like titanium and stainless steel.[30] As one can imagine after considering the discussion of stable neck propagation, the situation can be quite different for thermoplastics displaying the stable neck propagation phenomenon. Most importantly, the onset of localization need not be synonymous with ultimate failure. Unlike metals, thermoplastics do not automatically thin to failure at the initial localization site. The severe thinning that transpires when necking takes place may be unacceptable for particular applications, and therefore constitute failure. However, in spite of localization and thinning, a thermoplastic may still be able to undergo additional deformation and even additional load without ultimate failure because of the process of cold drawing and stable neck propagation.

Neck propagation and rupture in the puncture test

As mentioned in the last section, a very good example of how important the phenomenon of stable neck propagation can be with respect to the ultimate failure of thermoplastics can be found in the puncture, or Dynatup, test commonly used as a measure of impact resistance. Although this test is generally carried out at high rates of loading, the response of the specimen can often be quite similar at much slower rates. This test is often extremely nonlinear in nature. As a result, we will examine the role of these nonlinearities, especially those associated with large-strain behavior as they affect ultimate failure in such a test. Later, in Chap. 6, which deals with impact, some of the time-dependent effects of higher rate loadings will be considered.

The cross-sectional geometry of the test to be discussed was shown earlier in Fig. 5.21. The plastic disk shown in the figure is clamped between two metal plates with holes of radius a and loaded by a hemispherically tipped plunger of radius R. The behavior of this simple plastic disk under load will be analyzed using the finite-element technique. The two-dimensional, axisymmetric finite-element model used to carry out the numerical analysis is shown in Fig. 5.29. In the specific geometry considered, the unsupported radius of the disk, a, is 20 mm (0.8 in), the disk thickness, t, is 3.2 mm (0.125 in), and the radius of the hemispherical indenter, R, is 10 mm (0.4 in). As can be seen, the finite-element model includes the disk material clamped between the metal fixtures. This portion of the model becomes important as the displacements become large and significant yielding occurs near the supports.

There are a number of mechanical features that are important to the

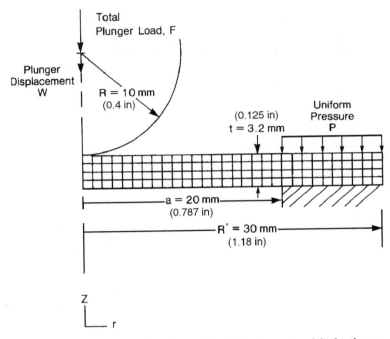

Figure 5.29 Two-dimensional, axisymmetric finite-element model of a thermo-plastic puncture test.

accurate modeling of this configuration. Reference 31 considers many of these issues in detail. For example, it is shown in this reference that frictionless boundary conditions at both the interface of the hemispherical head and plastic disk and the metal clamping rings at the outside boundary are representative of the actual behavior of the plastic disk in this test. In addition, the load application to the disk is modeled using a contact condition that defines the load distribution between load head and disk as part of the analysis. This is a nonlinear aspect of the problem mentioned in Chap. 3. At the support area, the plastic disk is prohibited from moving vertically by applying appropriate boundary conditions. However, the disk is allowed to move horizontally in an axisymmetric fashion as the indenter load is applied. Since the displacements incurred during the loading of this disk will be several times the thickness of the disk itself, large-rotation strain–displacement theory is required, as was discussed in the preceding chapter on stiffness. In addition, we will find that the strains in this problem become extremely large before failure, thus requiring the use of large-strain continuum theory and a material model that includes the nonlinear stress–strain relationship at these large strains. The numerical analysis is carried out using the ADINA finite-element program, as was the previously discussed analysis of the tensile specimen.

For this example, the disk will be considered to be made of polycarbonate. As in the case of the tensile specimen, an elastoplastic constitutive model is applied. Yielding is defined using the von Mises yield criteria. In addition, it will become important to account for the strain-hardening behavior observed in plastics such as polycarbonate, and the trilinear material model introduced in the discussion of the stable necking phenomenon observed in tensile tests will be used here also. A Young's modulus of 2.07 GPa (300 ksi) and a yield stress of 69 MPa (10 ksi) are used to describe the linear-elastic range of this polycarbonate polymer. The immediate post-yield modulus is approximated as zero. As part of the discussion, various values of the draw strain (ε_d) and the final hardening modulus (E_3) will be considered to illustrate the importance of these material properties to failure of this simple geometry. A more thorough discussion of these issues can be found in Ref. 32.

Figure 5.30 displays several load–displacement curves predicted with this numerical model using several variations of large-strain ma-

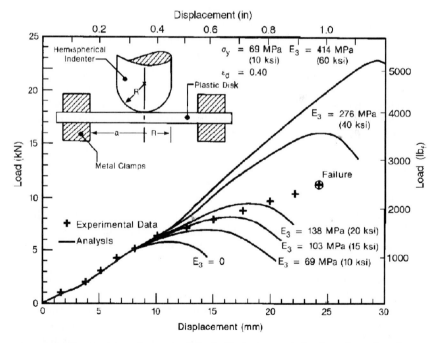

Figure 5.30 Comparison of measured load–displacement behavior of a polycarbonate puncture test with predicted response for various values of the final hardening modulus E_3 of the trilinear material model approximation.

terial properties and compares the results to measurements on an actual disk of polycarbonate. At crosshead displacements as low as 1–2 mm (0.04–0.08 in), the plastic disk under the indenter is predicted to begin yielding. In our previous discussion, it was suggested that this was one criterion that could be used to define damage such as permanent deformation. Through comparison with the experimental values, it is clear in this case, however, that there is significant load-carrying ability left in this polycarbonate disk before final failure. At a crosshead displacement of approximately 2.5 mm (0.1 in), the load–displacement curve of both the model and the experiment begins to stiffen and successively larger increments in load are required to produce equal increments of crosshead displacement. As was discussed in Chap. 3, this effect is associated with the increasing significance of the large-rotation terms in the nonlinear strain–displacement equations that occurs when plate displacements reach the order of the plate thickness—3.2 mm (0.125 in), in this case. The numerical model of this loading process using the simple elastic–perfectly-plastic constitutive model ($E_3 = 0.0$) continues to predict the load–displacement behavior of the disk very accurately until crosshead displacements as large as 10.0 mm (0.4 in) are reached. At this point, the agreement between model and experiment begins to disappear. The predicted load–displacement behavior indicates a maximum in its load–displacement curve in contrast to the experiment, which shows a reduction in the load–displacement slope but a continued capability to carry additional load. Consideration of the distorted shape of the model shown in Fig. 5.31 at the maximum of the predicted load–displacement curve associated with $E_3 = 0$ gives an indication of what is physically causing the maximum. As can be seen from Fig. 5.31c, the model of the plastic disk beneath the indenter is undergoing severe thinning at this point. This deformation is significantly different than what is visible at only a slightly earlier point in the test, also shown in Fig. 5.31b. This reduction in the thickness of the plastic disk is quite similar to the formation of the neck in the axisymmetric tensile specimen discussed previously. Because of the two-dimensional nonuniform stress state of the disk, yielding and necking do not take place simultaneously as they do in the simple tensile specimen. However, in a fashion quite similar to the tensile specimen, the rapid thinning process associated with the two-dimensional necking limits the load-carrying ability of the disk and creates a maximum in the load–displacement curve if the constitutive model after yield is perfectly plastic in nature ($E_3 = 0$). For a material described by this simple elastic–perfectly-plastic stress–strain relationship, additional crosshead displacement will lead to accelerated thinning at the center of the disk and continued load loss.

As was pointed out previously within the discussion of stable neck-

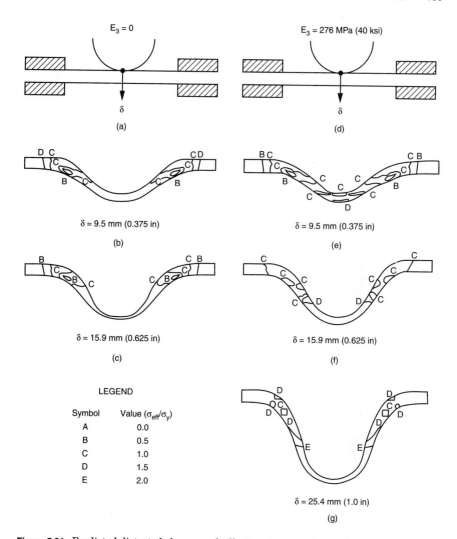

Figure 5.31 Predicted distorted shapes and effective stress contours of puncture test coupons for two different large-strain material models $(a–c)$ $E_2 = E_3 = 0$; $(d–g)$ $E_2 = 0$; $E_3 = 276$ Mpa (40 ksi).

ing observed in polycarbonate tensile tests, a simple elastic–perfectly-plastic constitutive model is not adequate in characterizing the behavior of polycarbonate at large strains. Polycarbonate actually displays an increasing slope in the true-stress versus true-strain relationship for strains on the order of 0.4 to 0.5. The effects of this type of material behavior on the puncture disk can be examined by using the trilinear stress–strain relationship employed to consider the behavior of the tensile specimen.

In a fashion parallel to our discussion of the effects of large-strain constitutive behavior on the tensile test, let us consider the effects of E_3 and ε_d on the load-carrying behavior of the puncture disk. First, let us assume that the value of ε_d is held constant at 0.4 while the value of E_3 is increased. Figure 5.30 illustrates load–displacement curves predicted for hypothetical materials with values of E_3 equal to 69 MPa (10 ksi), 103 MPa (15 ksi), 138 MPa (20 ksi), 276 MPa (40 ksi), and 414 MPa (60 ksi). These values represent ratios of E_3/σ_y of 1.0, 1.5, 2.0, 4.0, and 6.0, respectively. As can be seen in Fig. 5.30, any finite value of E_3 leads to the analytic prediction of an increased maximum load during a puncture test. This is a reflection of the fundamentally two-dimensional nature of the puncture test. The enhanced stiffness of the material in the vicinity of the indenter head serves to retard the thickness reduction at that location and force a redistribution of strain into surrounding areas of the disk, leading to significantly larger load maxima. For constant values of σ_y, E_2, and ε_d, larger values of E_3 lead to larger maximum loads as shown in the figure. The arrest of the thinning process and the subsequent propagation of the two-dimensional neck, which is caused by larger values of E_3, is clearly visible in Figs. 5.31d to 5.31g and parallels the behavior illustrated for tensile tests earlier. Using the simple trilinear stress–strain model, the model applied here eventually predicts the resumption of the thinning process beneath the indenter and an associated maximum in the load–displacement curve. It is unclear whether actual failure in the puncture disk is associated with this unstable thinning process or with some material-related parameter such as an ultimate strain or stress. However, it is certainly clear that the hardening process associated with finite values of the E_3 modulus in plastics has a significant effect on both the ultimate load and the total energy that can be absorbed by a thermoplastic under transverse load.

As was shown previously during the discussion of the tensile test, the amount of strain accumulated between the yield point and the point where hardening begins can also affect the load–displacement behavior of a thermoplastic. In order to examine this effect upon the puncture behavior, the value of E_3 in the trilinear constitutive model will be held constant at a value of 276 MPa (40 ksi) and the value of ε_d will be varied. Figures 5.32 and 5.33 illustrate the effects that the value of ε_d has upon the thickness reduction at the center of the disk and the disk's load–displacement behavior, respectively. As ε_d approaches infinity, the linear-elastic–perfectly-plastic material model becomes a limiting case for the trilinear model. For all values of ε_d, the thickness of the plastic sheet decreases monotonically during the punc-

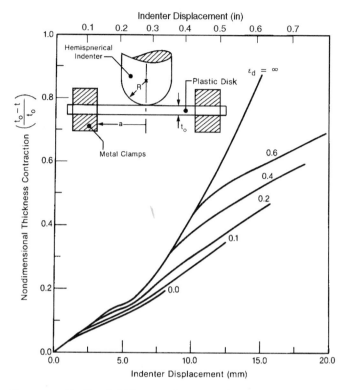

Figure 5.32 Predicted thickness reduction at the center of a puncture disk as a function of draw strain ε_d.

ture test. However, for the limiting case of ε_d approaching infinity, the rate of thickness reduction accelerates markedly at a crosshead displacement just over 5 mm (0.2 in). It is at this point that the two-dimensional necking process begins at the disk center. There is also a steady reduction in the slope of the load–displacement curve until eventually a maximum is reached at about 11 mm (0.43 in), as can be seen in Fig. 5.33. A significant difference in the predicted load–displacement behavior is visible in Fig. 5.33, if the value of ε_d is taken to be 0.6. Although the thinning behavior at the center of the disk is unchanged until a crosshead displacement of about 9 mm (0.35 in) is reached, the rate of thinning at the disk center is substantially reduced at that point. This reduction in the thinning process is a direct result of the material hardening that initiates when the true strain in the material reaches a value equal to $\varepsilon_y + \varepsilon_d$. In addition, since the disk now thins at a slower rate, crosshead loads of nearly twice the value associated with $\varepsilon_d = \infty$ are reached before the thinning process finally leads to

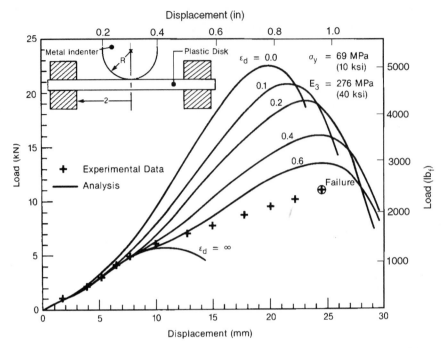

Figure 5.33 Predicted load–displacement behavior of a puncture disk as a function of draw strain ε_d.

a maximum in the modeled load–displacement curve shown in Fig. 5.33. As the value of ε_d is further reduced, it is evident from Fig. 5.32 that the change in the rate of disk thinning occurs at smaller and smaller values of crosshead displacement. As a result, the load–displacement curves associated with lower values of ε_d exhibit increased stiffness and larger maximum loads.

This model-based discussion makes it clear that the ultimate failure of ductile thermoplastic subjected to puncture loads is dependent upon the large-strain behavior of the plastic as well as its elastic modulus and yield strength. Prediction of performance associated with this type of failure requires the definition of stress–strain dependence at strains well beyond the yield point. However, measurement of stress–strain dependence for ductile thermoplastics at such large strains cannot be accomplished by normal test techniques using extensometers to measure strain because of the strain localization phenomenon discussed earlier. Alternative experimental approaches have been employed to define this behavior, including photographic techniques with standard American Society for Testing of Materials (ASTM) tensile bars[13] and axisymmetric tensile specimens with varying cross-sectional radii.[15]

Figure 5.34 presents large-strain data for polycarbonate from Ref. 13. As an indication of the accuracy in using the previously discussed trilinear constitutive model, the data in Fig. 5.34 was used to define σ_y, ε_d, and E_3; the model was then applied to predict the load–displacement behavior of a puncture disk with the same geometry as that shown in Fig. 5.29. The predicted load–displacement behavior is compared with experimental data in Fig. 5.35. As can be seen, the trilinear representation of the large-strain constitutive behavior of polycarbonate leads to a much more accurate prediction of the load–displacement performance of the disk than the simple-elastic–perfectly-plastic representation used previously. Figures 5.36a and 5.36b add additional evidence with respect to the ability of this modeling approach to predict physical behavior during the test. These figure panels compare the predicted cross-sectional shape of the 3.2 mm (0.125 in) thick disk at its maximum load with the actual permanently deformed shape of the punctured disk. Although the analytic shape in Fig. 5.36a is indicative of the shape at a crosshead displacement of 20 mm (0.8 in) and the picture of the disk in Fig. 5.36b is in its failed and unloaded state, the similarity of the cross sections is striking. The significant thickness reduction beneath the indenter head, which is predicted analytically, is clearly visible in the actual specimen. That reduced thickness is nearly constant over the region where the indenter contacts the disk in both analysis and experiment. Outside the region of contact, the specimen thickness approaches the original, undeformed value. The contour

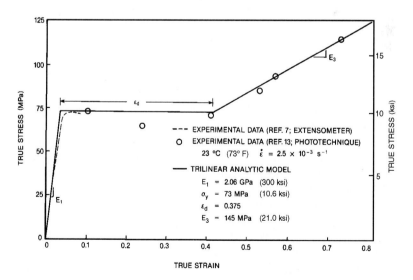

Figure 5.34 Trilinear, large-strain constitutive model fitted to stress–strain data for polycarbonate from Ref. 13.

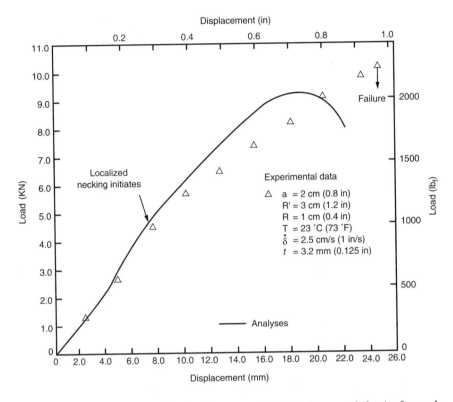

Figure 5.35 Comparison of predicted and measured load–displacement behavior for a polycarbonate disk subjected to puncture loads.

lines in Fig. 5.36a represent levels of nondimensional effective stress as predicted by analysis. The values of effective stress in the center of the disk where severe thinning has occurred are significantly above yield, indicating that the material has been worked well into the hardening range of Fig. 5.34. This material hardening, in turn, arrests the thinning process and forces the two-dimensional neck to propagate outward. Reference 33 presents several other examples illustrating the application of a trilinear, large-strain constitutive model to predicting large-strain ductile thermoplastic behavior.

The modeled results presented here provide insight into some of the fundamental material parameters that have a bearing on both the puncture resistance and formability of thermoplastic materials. In addition to Young's modulus and yield strength, the material hardening modulus E_3 observed in many ductile thermoplastic polymers after large strain, is shown to have a significant effect upon the predicted load–displacement behavior of a puncture test. This fundamental material property, which has been shown to be a controlling factor in the

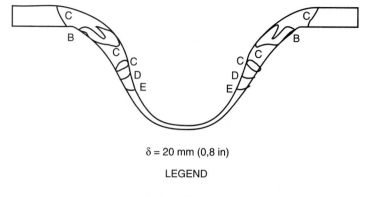

$\delta = 20$ mm (0,8 in)

LEGEND

Symbol	Value (σ_{eff}/σ_y)
A	0.0
B	0.5
C	1.0
D	1.5
E	2.0

(a)

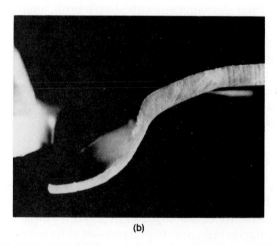

(b)

Figure 5.36 (*a*) Predicted deformed shape of a polycarbonate disk at maximum indenter load. (*b*) Cross-sectional view of actual ruptured disk.

stable necking process observed in tensile tests, can also lead to increased load-carrying capability in puncture tests. For larger values of E_3, the two-dimensional necking process in a puncture test is locally arrested and forced to propagate, thus providing the potential for increased energy absorption unless another failure mechanism intervenes. In addition, the value of the draw strain ε_d between yield and final hardening also plays an important role in puncture behavior. Larger values of ε_d result in more cross-sectional area reduction prior to the neck stabilization in both tensile and puncture tests. Thus, larger values of ε_d mean lower maximum loads before mechanical instability in the puncture test. If the values of E_3 and ε_d can be defined, current analysis techniques provide adequate predictions of load–displacement behavior in puncture tests.

In addition to understanding puncture resistance, the effect of these material properties, namely ε_d and E_3, is also significant to the process of cold-forming plastics. Unlike most metals, the existence of finite, increasingly hardening moduli at large strains, as observed in many polymers, makes necking possible without failure in a forming process. In situations where necking can be tolerated, the value of ε_d will govern the amount of "thinning" that will occur during forming. This will obviously have an effect upon final thickness.

The examples discussed in the last several sections have concentrated on failure associated with ductile behavior in polymers, including yield initiation as well as necking and rupture. In many respects, these are failures that are desirable insofar as they are reasonably stable with respect to their development and absorb large amounts of energy. Unfortunately ductile thermoplastics can also be susceptible to very brittlelike failure in circumstances often encountered in common engineering applications. Geometry can play a significant role in this change of failure modes. In some cases, a material exhibiting significant strain to failure and stable neck propagation in tensile and puncture tests may fail in a brittle manner at low strains in other geometries such as those associated with notches.

Notch Sensitivity and Crazing

Depending on the part geometry and associated stress state, some thermoplastics will exhibit both ductile and brittle failure modes. A good example of this behavior can be found in polycarbonate. For stress states that are primarily one- and two-dimensional in nature such as tensile and puncture tests, polycarbonate remains extremely ductile at temperatures as low as -80 to $-100°C$ (-112 to $-148°F$).[34] However, if the geometry of the part is such that a significant three-dimensional stress state is generated, such as in the case of a notched beam, then polycarbonate can fail in a very brittle fashion at room temperature.[35] This latter type of behavior is often referred to as notch sensitivity. It is an important phenomenon because the brittle failure of a normally ductile material can be a very catastrophic event. Unfortunately, failure due to notch sensitivity is neither as well understood nor as predictable a failure event as is yielding. However, understanding has progressed to the extent that general geometries susceptible to brittle failure can be identified. Furthermore, specific geometric parameters can also be assessed with respect to the goal of eliminating brittle failure.

In order to build some understanding of notch-sensitivity failures in thermoplastics, it is important to have an understanding of another damage mode characteristic of these materials, namely, crazing. Amorphous thermoplastic polymers (and to some extent semicrystalline thermoplastics also) display a phenomenon known as *crazing*. Although similar at first glance to a crack in some respects, the detailed structure of a craze is significantly different. Specifically, the morphology of a craze includes fibrils of highly oriented material stretched between free surfaces. These fibrils provide load-carrying ability between the crack surfaces. This peculiar thermoplastic phenomena has been intensely studied since 1949, and general discussion of it as well as its causes and consequences can be found in the literature.[36–38] Specifically of interest in the present investigation is the relationship between crazes and brittle failure.

Crazing is a process whose initiation is affected by numerous factors including the presence of solvents, molding conditions, annealed state, and stress states characterized by large positive values of the first stress invariant (hydrostatic stress). It is also a time- and temperature-dependent process. The primary focus of this present discussion will be on the stress state and associated geometry dependence of craze formation. From a structural design point of view, crazes are significant because they represent a potential for brittle, catastrophic failure.

In order to assess whether a particular material, geometry, and load condition will produce damage in the form of a craze, a criterion is necessary. Several criteria have been suggested.[39–42] The data that serve as the basis of one of these criteria[39,40] were measured for polymethylmethacrylate (PMMA). Although limited to plane-stress conditions, the experiments are quite systematic and well controlled. In two sets of experiments, Sternstein and Ongchin[39] and Sternstein and Myers[40] studied the appearance of crazing and yielding in PMMA under a two-dimensional stress field. Their observations seem to indicate that high values of the first invariant of the stress tensor (hydrostatic tension) enhance the formation of crazes. On the other hand, hydrostatic compression tends to inhibit craze formation and make yielding more likely. As a criterion for crazing that adequately fits their data, Sternstein and Ongchin propose

$$| \sigma_1 - \sigma_2 | = A(T) + B(T)/I_1 \qquad (5.4)$$

where $A(Y)$ and $B(T)$ are temperature-dependent constants and I_1 is the first invariant of the stress tensor. They show that, in general, the curves representing craze initiation and yielding in PMMA may intersect, as shown in Fig. 5.37. When they do intersect, crazing will occur for some stress states, while yielding will occur for others.

Since both the crazing and the yielding phenomena are temperature sensitive, the relative positions of the two surfaces may change as a function of temperature. As a result, for one temperature, crazing may be the active damage mode in the first quadrant of the plane-stress field while at other temperatures, yielding may prevail. This is, of course, also true with respect to different materials at the same temperature. Furthermore, for some materials and conditions, the criteria defining craze initiation may fall completely outside the yield surface for all plane-stress states. In such cases, yielding will occur first for any plane-stress condition. As pointed out previously, Yee and Carapellucci[6] showed that this was the case for polycarbonate at room temperature.

So far the discussion of damage criteria has been limited to the plane-stress state. Obviously there are some geometry and load combinations where the stress state is three-dimensional in nature. In this situation the same question relative to which damage mode is active can be posed. One simple geometry that produces a three-dimensional stress state is a beam with a notch on one of the surfaces as shown in Fig. 5.38. This notched geometry is a characteristic part of both the Charpy and Izod impact tests that will be discussed later in Chap. 6. Depending upon the details of the local geometry of the notch in this beam, the failure mode of the beam under a centrally concentrated load can vary significantly. Figure 5.39 illustrates some of the important characteristics of these differences in failure for a polycarbonate beam as a function of the notch root radius. If there is no notch in the beam, the failure of the beam is dominated by large-scale plastic flow, as can

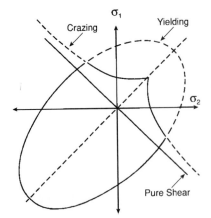

Figure 5.37 Biaxial yield and craze limits (Sternstein and Ongchin, Ref. 39).

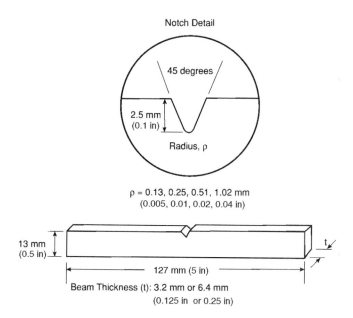

Notch Detail

45 degrees

2.5 mm
(0.1 in)

Radius, ρ

ρ = 0.13, 0.25, 0.51, 1.02 mm
(0.005, 0.01, 0.02, 0.04 in)

13 mm
(0.5 in)

127 mm (5 in)

t

Beam Thickness (t): 3.2 mm or 6.4 mm
(0.125 in or 0.25 in)

Figure 5.38 Notched beam geometry.

be seen by comparing the undeformed beam at the top of the photograph of the figure with an unnotched beam just below it, which has been deformed beyond maximum load. Although a limit load associated with a fully plastic hinge beneath the concentrated load is reached, large displacements can be incurred while the beam remains intact, as shown in the figure. If a notch with a 1.02 mm (0.04 in) radius is placed on the bottom surface of the beam, the deformation and failure behaviors of the beam remain dominated by plastic flow, as can be seen from the third specimen in Fig. 5.39. The beam eventually tears at the root of the notch but only after significant plastic deformation. The load–displacement curve for this geometry, also shown in Fig. 5.39, displays significant additional displacement after reaching a maximum load. However, if the notch radius is decreased to 0.13 mm (0.005 in), the failure of the beam is distinctly different. First of all, the beam suddenly and catastrophically loses its load-carrying ability while still in a nearly linear portion of its load–displacement behavior. Second, this notch geometry leads to complete fracture of the beam into two pieces, as shown in the fourth specimen at the bottom of the figure. Furthermore, the beam fracture surfaces are very glassy with little evidence of macroscopic plasticity. From an engineering point of view, avoiding this latter mode of failure is clearly desirable.

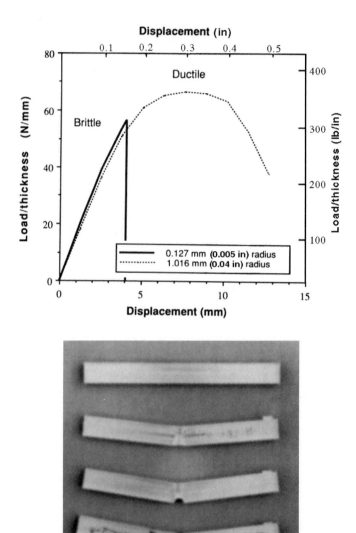

Figure 5.39 Different failure modes visible in a notched polycarbonate beam as a function of notch-tip radius.

There is evidence that the same basic phenomena of yielding and crazing that define alternative failure modes in two-dimensional stress states are also present in three-dimensional stress states like those beneath a notch. Crazes have been observed to appear beneath notches. Although experiments relevant to crazing in these geometries are much more complex and of necessity not as well controlled as the two-dimensional tests of Sternstein, Ongchin, and Myers, numerous investigations have been reported in the literature.[35,43–50] Many of these experiments have studied polycarbonate because its transparent optical properties allow the craze event to be observed. Reference 35 presents similar data for polyetherimide, which is also optically transparent.

One of the most interesting characteristics of craze formation in these notched geometries is the fact that the crazes are observed to initiate subsurface to the notch. Understanding the stress state beneath the notch in a simply supported beam helps to explain the subsurface location of the craze appearance. The solution for the stress state in the vicinity of the notch based upon slip-line field theory[51]—often used in support of the experiments using notched geometries—characterizes some of the significant aspects of this stress field. First, although the maximum principal stress and the maximum hydrostatic stress are located at the notch-tip surface during elastic response of the beam, as plasticity begins to occur at the notch tip, the location of maximum stress shifts to the elastic–plastic boundary just below the notch-tip surface. For a material with an octahedral shear yield stress τ_0 and a notch-tip radius ρ, successively larger values of hydrostatic pressure and the maximum principal stress are reached as the elastic–plastic boundary is forced farther away from the notch tip. Within the context of slip-line field theory, the maximum hydrostatic pressure p and the maximum principal stress σ_1 are given by

$$P = \tau_0[1 + 2\ln(1 + x/\rho)] \qquad (5.5)$$

$$\sigma_1 = 2\tau_0[1 + \ln(1 + x/\rho)] \qquad (5.6)$$

where x is the distance between the notch-tip surface and the elastic–plastic boundary. More recently, elastic–plastic finite-element solutions of this notched-beam problem[35,52] have also been carried out as a means of relaxing some of the assumptions of rigid-plastic constitutive behavior and small-strain kinematics inherent in the common slip-line field solutions. Figures 5.40a and 5.40b provide analytically predicted evidence of the location of the largest values of maximum principal stress before and after onset of yielding at the notch root.[35] Hydrostatic stress maxima occur at the same location. Careful microscopic examination of the material below the notch surface[35,43–50] indicates that when a craze appears in this geometry, it appears at the boundary of the elastic and plastic zones where the hydrostatic stress and principal stress are maximum.

For a material such as polycarbonate that yields rather than crazes in a benign chemical environment at moderate loading rates, the two-dimensional stress state imposed by the free surface of the notch ensures that yielding will occur there first, as shown in Fig. 4.50a.

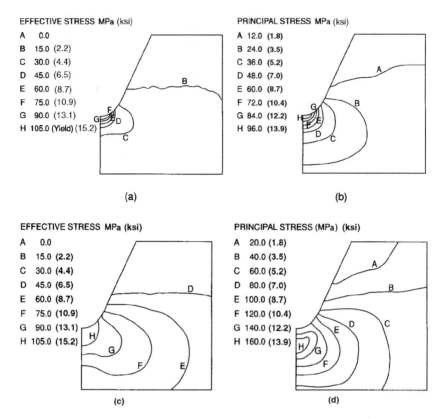

Figure 5.40 Distribution of (a) effective stress and (b) maximum stress at a load prior to initial yield; distribution of (c) effective stress and (d) maximum principal stress at a load after initial yielding.

However, as the plastic zone beneath the notch expands, as shown in Fig. 5.40c, and the values of principal stress and hydrostatic tension at the elastic–plastic boundary increase, crazing eventually becomes possible.

When a craze does occur beneath the notch surface, it has an important effect upon the character of failure in the beam. Reference 35 reports experimental results for notched beams made of polycarbonate and polyetherimide loaded in three-point bending. Both of these materials yield and draw to strains on the order of 50% in tensile tests. Four notch root radii [0.13, 0.25, 0.51, 1.02 mm (0.005, 0.01, 0.02, 0.04 in)] and two beam thicknesses [3.2 and 6.4 mm (0.125 and 0.25 in)] were tested. In the case of the polyetherimide, all the beams failed in a brittle fashion except the 3.2 mm (0.125 in) thick beams with 1.02 mm (0.25 in) notch radii. In every case, the appearance of a craze was observed to precede the final catastrophic failure of the beam. In the one set of beams that did not fail in a brittle fashion, no crazes were observed to form and the beams underwent large-scale plasticity with tearing failure finally initiating from the notch surface. In the case of the polycarbonate, crazes were only observed to form in two geometries at room temperature—6.4 mm (0.25 in) thick beams with 0.13 mm (0.005 in) notch-tip radii and 6.4 mm (0.25 in) thick beams with 0.25 mm (0.01 in) notch-tip radii. The beams with the 0.13 mm (0.005 in) radii failed in a brittle manner after the craze event. In the case of the beams with the 0.25 mm (0.01 in) notch-tip radii, the failures were not brittle in nature but rather associated with tearing from the notch tip. None of the other polycarbonate geometries showed any evidence of crazing. Instead, they all failed after large-scale plasticity and tearing.

These observations from Ref. 35 are consistent with results from finite-element analyses carried out as part of the same investigation that focused on the importance of maximum principal stress and hydrostatic stress to craze initiation as suggested by earlier work.[39,42,50] One of the significant results of these analyses, which made use of elastoplastic constitutive models and large-strain continuum mechanics, is that there are geometrically associated limits to the values of maximum principal stress and hydrostatic tension that can occur beneath the notch of the beam loaded in three-point bending. These limits are dependent upon the values of the beam thickness and the notch-tip radii and are successively lower for beams with smaller thicknesses and larger notch-tip radii. Although an explicit failure criterion is not suggested in Ref. 35, the results of the numerical modeling and the experimental observations show that maximum principal stress and hydrostatic tension are important to the craze initiation event in notched geometries just as they are in two-dimensional stress states. The continuum mechanics analyses in Ref. 35 shows that reducing the beam thickness or increasing the notch-tip radius will lower the limiting values of both maximum principal stress and maximum hydrostatic stress attainable during a flexural test, as shown in Figs. 5.41 and 5.42. If the limiting values associated with the maxima in Figs. 5.41 and 5.42 are sufficiently reduced through manipulation of these geometric quantities, then crazing will not occur. If crazing is precluded, then the experimental observations discussed earlier suggest that the failure will be nonbrittle in nature.

In Ref. 35, the authors assess the consistency of applying criteria based upon maximum principal stress and maximum hydrostatic stress for prediction of craze appearance in polycarbonate and polyetherimide beams with various notch-tip radii and thicknesses. Such criteria offer the advantage of being relatively simple to apply. Figure 5.43 is one example of their comparisons using the hydrostatic stress criteria for polycarbonate. The results were consistent in that critical values of hydrostatic stress that could be inferred as necessary to pro-

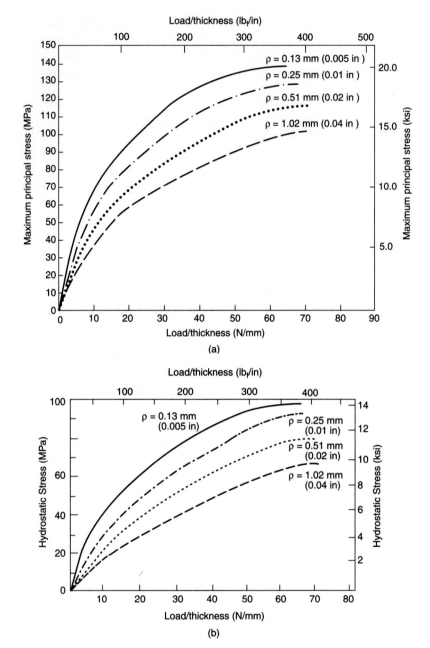

Figure 5.41 Predicted effect of notch-tip radius (ρ) upon (a) maximum principal stress and (b) maximum hydrostatic stress at the plastic zone boundary in a 6.4 mm (0.25 in) thick polycarbonate beam.

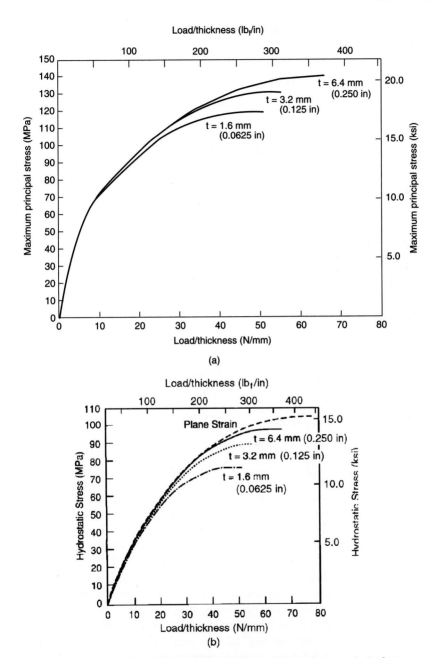

Figure 5.42 Predicted effect of beam thickness (t) upon (a) maximum principal stress and (b) maximum hydrostatic stress at the plastic zone boundary in a polycarbonate beam with a 0.13 mm (0.005 in) notch-tip radius.

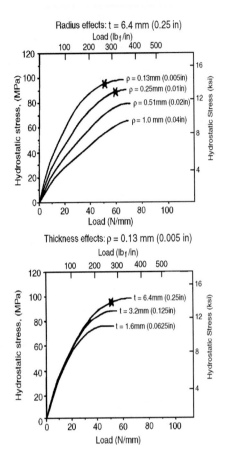

Figure 5.43 Comparison of predictions for maximum attainable hydrostatic stress and predictions of hydrostatic stresses associated with craze appearance in experiments using polycarbonate beams.

duce crazes in beams with notch-tip radii of 0.13 mm (0.005 in) and 0.25 mm (0.01 in) were higher than the analytically predicted maximum attainable values for hydrostatic stresses in notch geometries with radii of 0.51 mm (0.02 in) and 1.02 mm (0.04 in) where no crazing occurred. Results associated with a maximum principal stress criteria were very similar. However, there was a lack of internal consistency in results for the two smallest notch-tip radii. Since there were only a limited number of test results in this study, it is difficult to assess the effects of statistical scatter on the consistency of prediction. Results for polyetherimide presented in Ref. 35 are qualitatively similar but exhibit more scatter in inferred critical stress values.

Reference 53 presents a much more statistically representative study of polyetherimide, using analyses and tests similar to Ref. 35. For each geometric configuration, 15 to 25 test specimens were tested. After assessing their results, the authors found that the criteria, suggested in Ref. 41,

$$\sigma_1 - \upsilon(\sigma_2 + \sigma_3) = \frac{X(T)}{\sigma_1 + \sigma_2 + \sigma_3} + Y(T) \qquad (5.7)$$

where υ is Poisson's ratio, $X(T)$ and $Y(T)$ are temperature-dependent constants, and σ_1, σ_2, and σ_3 are principal stresses at a point, provided a very consistent criteria for craze initiation over all geometries when used in conjunction with a pressure-sensitive yield criterion defined by

$$\tau_y = \tau_0 - \mu P \qquad (5.8)$$

where τ_y is the shear yield stress, P is the hydrostatic stress, τ_0 is the shear yield stress corresponding to zero hydrostatic stress, and μ is a material constant.

There are numerous common plastic shapes that lead to stress states quite similar to the notched beam. A ribbed plate in bending, such as the one shown in Fig. 5.44, can produce a stress state and stress values quite similar to the notched beam depending on the details of its local geometry. Currently it is still very difficult to predict whether an arbitrary geometry will produce local three-dimensional stress states consistent with the craze event. The numerical modeling carried out in Refs. 35, 52, and 53 is extremely detailed in nature, as illustrated by the dense three-dimensional finite-element mesh used for the notched beam analyses and shown in Fig. 5.45. This type of analysis is not consistent with standard engineering design practices for plastic parts. Furthermore, neither definitive stress-based criteria for crazing nor standardized testing techniques to establish those criteria as a function of temperature exist. However, the current level of understanding does lead one to hope that design tables encompassing general geometric shapes could eventually be developed as an aid to engineers for defining general geometric shapes susceptible to brittle failure as well as specific geometric parameters that have the most effect upon brittle failure in a particular geometry.

Temperature Effects upon Failure Mode

As is clear from the discussion in the last two sections, geometry and the associated stress state can play a significant role in determining

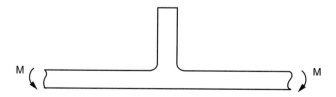

Figure 5.44 Ribbed plastic plate in bending.

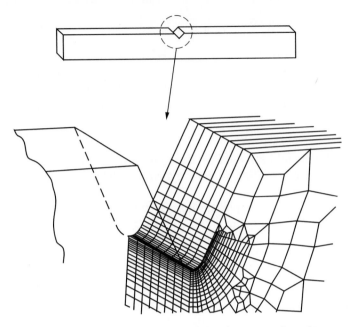

Figure 5.45 Typical three-dimensional finite-element mesh used to analyze notched-beam behavior.

the failure mode of a plastic component. Some thermoplastics such as polycarbonate and polyetherimide may exhibit extreme ductility in simple geometries like a tensile bar or flat plate, but may fail in a brittle fashion in other geometries like the notched beam. In addition to geometry, temperature can also play a major role in determining the failure mode of a plastic component.

As an example, consider the behavior of a polycarbonate disk subjected to a centrally located indenter load that was previously discussed in this chapter. Both the experiments and the analyses discussed in that section were relevant to room-temperature behavior and the final failure of the disk, whose load–displacement relationship is shown in the upper portion of Fig. 5.46, was characterized by ductility visible in Fig. 4.57a and extremely large strains (~40 to 50%). However, if the temperature of the thermoplastic in that test is reduced to –90°C (–130°F), the plate fails at much lower loads (and associated strains) as can also be seen in Fig. 5.46 and in an extremely brittle fashion as shown in Fig. 5.47b.

Unfortunately, transition temperatures from ductile to brittle failure do not appear to be material properties. Geometry and the associated stress state can have a significant effect upon this transition temperature from ductile to brittle behavior. For multiaxial states of stress defined by the three principle stress components σ_1, σ_2, and σ_3, the

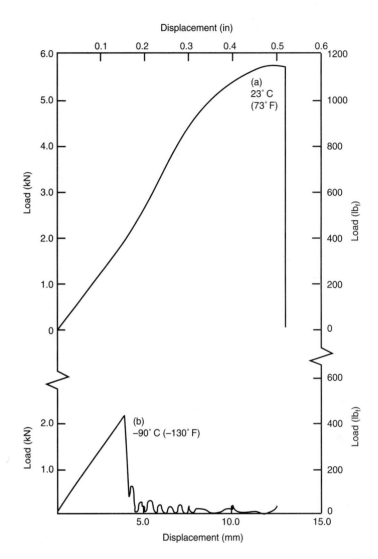

Figure 5.46 Experimentally observed maximum load for a centrally loaded polycarbonate disk at (a) room temperature and (b) –90°C (–130°F).

(a)

Figure 5.47 Failed polycarbonate disks tested at (a) room temperature and (b) –90°C (–130°F).

character of the stress state can be described mechanically as having a dilatational (volumetric) component that can be quantified with the hydrostatic stress defined as

$$P = \tfrac{1}{3}(\sigma_1 + \sigma_2 + \sigma_3) \qquad (5.9)$$

and a deviatoric (shear) component quantified by the octahedral shear stress

$$\tau_0 = \tfrac{1}{3}\sqrt{(\sigma_1 - \sigma_2)^2 + (\sigma_2 - \sigma_3)^2 + (\sigma_3 - \sigma_1)^2} \qquad (5.10)$$

States of stress in which the ratio of hydrostatic stress to octahedral shear stress is positive and large are generally associated with higher transition temperatures than stress states with a lower ratio, dominated by shear deformation. This effect can be illustrated in several ways. Reference 34, for example, shows that a polycarbonate material subjected to a biaxial state of stress in a grooved tensile bar, shown in

(b)

Figure 5.47 (*Continued*)

Fig. 5.48c $(P/\tau_0 \approx 0.8)$ is associated with a higher transition temperature than that associated with a tensile test $(P/\tau_0 \approx 0.408)$. Still greater
values of the ratio of hydrostatic to deviatoric stress are achieved in a
beam specimen with a notch, as shown in Fig. 5.48d. For this geometry,
the transition temperature from ductile to brittle failure is still higher
and dependent upon the exact geometry of the notch. The net practical
result of these effects is that the likelihood of a brittle fracture of a
plastic component is much higher at lower temperatures, and for
highly constrained (thick) plastic components in the presence of
notches and similar stress concentrators. Higher values of strain rate
also have the effect of promoting brittle failure, and this dependence
will be discussed more in Chap. 6 with regard to the subject of impact.

Reference 54 summarizes work reported in this area with respect to
plastics and notes some of the observed similarities in behavior with
respect to other materials. Pragmatically, the definition of a ductile-to-
brittle transition temperature would seem to be an extremely important engineering variable. However, there does not appear to be an
agreed-upon method of measuring and reporting such a quantity in a

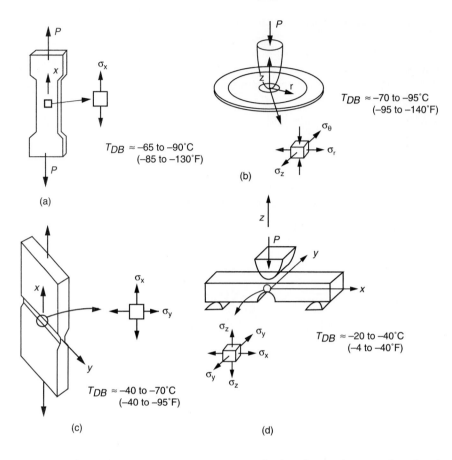

Figure 5.48 Approximate transition temperatures of polycarbonate from ductile to brittle failure for several different geometries: (*a*) tensile bar, (*b*) puncture disk, (*c*) grooved tensile bar, and (*d*) notched beam.

manner that is easily applicable for an engineer dealing with general geometric configurations. As an initial aid to those involved in material choice, Vincent[55] suggests that plastics be considered with respect to three categories:

1. Brittle even when unnotched

2. Brittle when notched

3. Tough; specimens do not break completely even when sharply notched

TABLE 5.3 Failure Characteristics of Thermoplastics as a Function of Temperature*

	Temperature							
	−20	−10	0	10	20	30	40	50	°C
	−4	14	32	50	68	86	104	122	°F
Polystyrene	A	A	A	A	A	A	A	A
Polymethyl methacrylate	A	A	A	A	A	A	A	A
Glass-filled nylon (dry)	A	A	A	A	A	A	A	B
Polypropylene	A	A	A	A	B	B	B	B
Polyethylene terephthalate (PET)	B	B	B	B	B	B	B	B
Acetal	B	B	B	B	B	B	B	B
Nylon (dry)	B	B	B	B	B	B	B	B
Polysulphone	B	B	B	B	B	B	B	B
High-density polyethylene (HDPE)	B	B	B	B	B	B	B	B
Rigid polyvinyl chloride (PVC)	B	B	B	B	B	B	C	C
Polyphenylene oxide	B	B	B	B	B	B	C	C
Acrylonitrile-butadiene-styrene (ABS)	B	B	B	B	B	B	C	C
Polycarbonate	B	B	B	B	C	C	C	C
Nylon (wet)	B	B	B	C	C	C	C	C
Polytetrafluoroethylene (PTFE)	B	C	C	C	C	C	C	C
Low-density polyethylene (LDPE)	C	C	C	C	C	C	C	C

*A: Brittle even when unnotched; B: notch brittle; C: tough; specimens do not break completely even when sharply notched.

As we have seen, some materials will display more than one of these behavior types as a function of temperature. Therefore, Vincent[55] has constructed a table of behavior for plastics as a function of temperature, which is shown in Table 5.3. After assessing the product's design requirements, an engineer could use such a table to guide initial material choices.

Fracture Mechanics

Having considered the effect that notch sensitivity has with respect to brittle failure in normally ductile thermoplastics, it should not be difficult to understand that a crack in a thermoplastic part can have an equally deleterious effect. Figure 5.49 graphically illustrates the difference that the presence of a crack can make on the load-carrying ability and failure of a thermoplastic part. In this example the temperature was −50°C (−58°F). The general test geometry shown in Fig. 5.49 is similar in many respects to the box section, the failure of which was previously considered as part of the discussion on yielding and plasticity. The only geometric differences are the absence of the backplate of

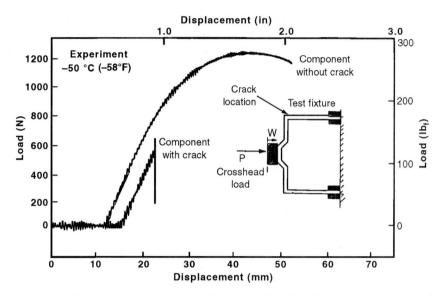

Figure 5.49 Comparison of experimentally measured load–deflection curves [at –50°C (–58°F)] representative of a U-channel with and without a crack.

the box section and the associated changes in the support fixturing. In this case, the load was applied at the front of the box in a fashion quite similar to the previous example. The U-shaped channel in Fig. 5.49 was made from a thermoplastic blend of polycarbonate and polybutylene terephthalate. The stress–strain behavior of this material is very similar to that of polycarbonate, including its large-strain behavior. Also shown in Fig. 5.49 are two measured load–displacement curves. The only physical difference between the two tests was the presence of a crack in the upper side of one of the two U-channels. The very linear load–displacement curve that terminates with fracture and a sudden and total loss of load-carrying ability is associated with the cracked section, whereas the very nonlinear curve is associated with test results at the same rate and temperature for a part without a crack. In contrast to the brittle, catastrophic failure of the cracked section, the channel section without a crack experienced a maximum load due to yielding and slow reduction in load-carrying ability, which is very similar to the response of the rubber-toughened polycarbonate box tested at room temperature and discussed earlier in this chapter. The uncracked section did not fracture.

Quantitative treatment of failure in the presence of cracks like this requires the application of fracture mechanics techniques. The geome-

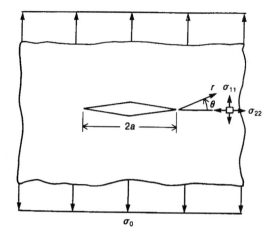

Figure 5.50 Crack in a tensile stress field.

try of primary interest in this case is a crack subjected to a tensile stress field shown in Fig. 5.50. For such a geometry, the stresses perpendicular to the crack approach infinity at the crack tip. Their variation as a function of distance from the crack tip is expressed as

$$\sigma_{11} = K_I/\sqrt{2\pi r} \tag{5.11}$$

where K_I is a function of the applied load as well as the component and crack geometry and is referred to as a stress-intensity factor. Obviously stress can no longer be used as a failure criteria at the crack tip, since the value of stress in Eq. (5.11) goes to infinity. However, it has been shown that under the proper circumstances there is a true material property that can be used to predict failure quantitatively. The property used in these situations is the critical stress-intensity factor (K_{Ic}), also referred to as the plane strain fracture toughness.

The value of the critical stress intensity of a material can be measured by testing standard cracked specimens, such as the compact tension (CT) specimen shown in Fig. 5.51. For the compact tension test, the stress-intensity factor K_I can be related[56] to the test load P as

$$K_I = (P/BW^{1/2})F_{CT}(a/W) \tag{5.12a}$$

where B and W are the compact specimen thickness and width defined in Fig. 5.51, a is the crack length, and

$$F_{CT}(x) = [(2+x)(0.866 + 4.64x - 13.32x^2 + 14.72x^3 - 5.6x^4)/(1-x)^{3/2}] \tag{5.12b}$$

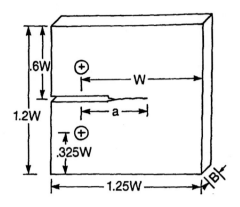

Figure 5.51 Compact tension specimen.

$$x = a/W \tag{5.12c}$$

Using Eq. (5.12a), a relationship between the load at failure, P_F, and the stress-intensity factor associated with that failure load, K_Q can be established:

$$K_Q = (P_F/BW^{1/2})F_{CT}(a/W) \tag{5.13}$$

If the fracture event meets appropriate standards for applicability of linear-elastic fracture mechanics (LEFM), defined in Ref. 56, then the value of K_Q is defined as the plane-strain fracture toughness or critical stress intensity factor K_{Ic} and represents a material property.

As a material property, K_{Ic} can then be used to quantify failure of components with general geometric configurations. In order to predict the fracture load of a cracked structure, the stress intensity factor for the part and the applied load must be identified. Stress intensity factors for some generic configurations have been tabulated in handbooks like Ref. 57. For example, the stress-intensity factor for a simply supported, centrally loaded beam with a surface crack of length a on the beam face opposite the load, shown in Fig. 5.52 is

$$K_I = \sigma\sqrt{\pi a}\,F_B(a/h) \tag{5.14}$$

where

$$\sigma = 3PL/2h^2 \tag{5.15}$$

where L and h are the beam span and depth, respectively, and P is the force per unit width of the beam. For the specific geometric configuration defined by $L/h = 8$, $F_B(a/h)$ is defined as

and experimental measurements in a low-temperature fracture event as reported in Ref. 67. The component tests considered were channel sections previously illustrated in Fig. 5.49 and made of the same rubber-toughened polycarbonate–polybutylene terephthalate blend with a temperature-dependent fracture behavior shown in Fig. 5.53. The component tests were conducted well below the transition temperature at $-50°C$ $(-58°F)$ and at the same stress-intensity rate as the compact tension tests used to define K_{Ic} [180 MPa$\cdot$ m$^{1/2}$ (165 ksi$\cdot$ in$^{1/2}$/s)].

The solid lines in Fig. 5.58 represent the bounds on the analytic prediction as a result of uncertainty in boundary conditions. The uncertainty in this problem arose primarily from the fact that the top and

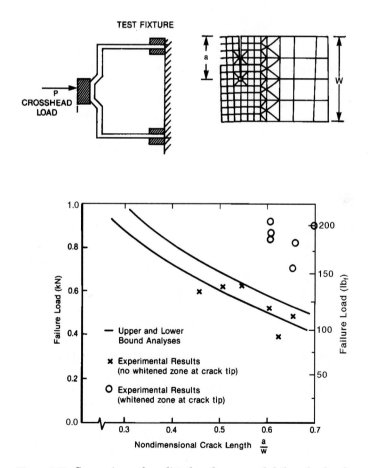

Figure 5.58 Comparison of predicted and measured failure loads of a cracked thermoplastic component.

bottom portions of the channel did not always form a 90° angle with respect to the front portion of the box. As a result, when the channel was placed in the mounting fixture, there was some preloading of the crack present. This preloading varied from piece to piece. Measurements of a cross section of the tested geometries were taken and their extreme values were used to calculate the bounding lines in the figure.

Although simple closed-form expressions are tabulated for some simple geometries like the beam discussed earlier, in the case of more general geometries, an analytic expression for the stress-intensity factor may not always be available or accurate. There are several analytic techniques available for the prediction of stress-intensity factors in such general engineering components.[58-61] For the analysis discussed here, the finite-element technique was applied with crack-tip singularity elements (Refs. 60 and 61) used to characterize the stress field in vicinity of the crack tip. This technique is based upon the fact that quadratic, isoparametric finite elements with midside nodes placed at quarter-point locations in elements defining the crack tip will exactly reproduce the $1/\sqrt{r}$ stress singularity at the crack tip. The nodal displacements in these elements can then be used to calculate the stress-intensity factor describing the cracked structural configurations. The important point to be understood here is that numerical techniques are available to apply in more general geometry configurations. For details of this particular approach, the reader is directed to the references listed above.

The finite-element mesh in the vicinity of the crack that was used to analyze this test is also shown in Fig. 5.58. The crack in that figure extends halfway through the component wall. In order to approximate the curves in Fig. 5.58, which are presented as functions of the crack length, cracks extending one-eighth and one-quarter of the way through the thickness were also modeled. Eight-node, two-dimensional, isoparametric elements were used except in the vicinity of the crack tip where six-node triangular elements were employed.

The cracks in the component were also introduced in fatigue from a razor notch as they were in the compact tension specimen tests. When sharp cracks were achieved (the circles in Fig. 5.58), the test results agreed well with prediction. When blunt cracks appeared along with stress-whitening observed after fracture at the crack tip, higher fracture loads were noted. This is consistent with the concept of LEFM as a lower-bound failure prediction. The agreement between analysis and experiment is very good. This example illustrates the fact that when a crack length can be identified and carefully measured fracture toughness data are available, LEFM can be effectively used to predict conservative failure loads in general geometries.

As a final remark with respect to the engineering application of fracture mechanics for failure prediction, it is worth reemphasizing a fundamental difference between fracture mechanics methodology and other failure methodologies discussed in this chapter. Although plane strain fracture toughness is a material property, knowledge of its value alone is not sufficient to predict a structure's failure load. In addition to K_{Ic}, one must also be able to quantify the length of the crack in the structure. Crack lengths are obviously not a design parameter for structures. They represent a measure of accumulated damage. As a result, in order for fracture mechanics and K_{Ic} properties to be applicable in a design process, some methodology to define crack size as a function of service lifetime is necessary. For other engineering materials, fracture mechanics has been applied in conjunction with nondestructive testing techniques to identify crack sizes at specific lifetime intervals of a structure. If a sufficiently large crack is identified in a critical structural component, it can then be removed from service on the basis of fracture mechanics. The fact that unfilled thermoplastics are not commonly used in structures so critical as to warrant nondestructive testing at specific intervals of lifetime limits the engineering application of such data for thermoplastic structures.

Instability and Collapse of Thin Plastic Components under Compression

The previous discussions with respect to failure of thermoplastic parts with either stress concentrations or cracks focus primarily upon failure events driven by tensile stress fields. Failure due to compressive stresses is also a significant concern in plastic parts. Of course, a plastic part can yield in compression and fail just as easily as it does in tension. The general elastoplastic constitutive relationships available in most finite-element codes will account for this possibility in a fashion identical to one dominated by tensile stresses. However, there is an additional mode of failure that must be considered carefully for plastic parts that is driven by compressive stresses—namely, buckling instability and subsequent structural collapse. A significant difference between this type of failure and those discussed previously is the primary role that structural geometry plays in the definition of this failure in contrast to the material failures discussed in all the previous examples in this chapter.

As was pointed out in the previous chapter on stiffness, plastics have Young's moduli that are often two orders of magnitude less than steel. In situations where structural stiffness is a design issue, the shape of the plastic part must often be used to help offset this significant difference in material stiffness. Consequently, an attractive cross-sectional shape for a thermoplastic beam is a box section because of its effectiveness in producing bending stiffness. However, as the moment of inertia of the box beam is increased within the constraints of wall thicknesses that are realistic for the injection-molding process, a plastic beam begins to behave more and more like a thin shell structure than a simple beam. The effect of this trend on beam stiffness was already explored in the preceding chapter. In some situations, large deformation in shell-like plastic parts can actually lead to failure due to structural instability.[69] In the example that we will discuss next, the generic thermoplastic box beam introduced in the last chapter will be reexamined with emphasis on this type of failure. As in the preceding chapter, finite-element analysis will be used to examine these nonlinear structural issues. The insets in Fig. 5.59 define the cross-sectional geometry of the section under consideration and the boundary conditions at the

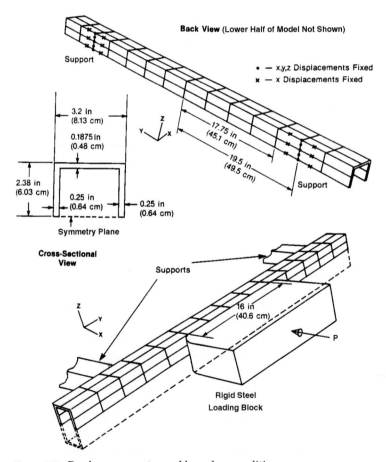

Figure 5.59   Box-beam geometry and boundary conditions.

support areas. The loading in this investigation is applied by a 40 cm (16 in) rigid steel block at the center of the beam. This structure is obviously similar to a bumper beam, many of which are now made of thermoplastic material. Details of the investigation can be found in Ref. 69.

To begin, Fig. 5.60 compares the load–displacement behavior as predicted by nonlinear analysis techniques for several different beam configurations and helps illustrate the importance of several structural instability events contributing to significant nonlinearity and potential failure of the beam. For reference, the linear solution to the flexurally loaded, thin-walled box beam is also plotted in Fig. 5.60. As can be seen in the figure, the nonlinear analysis of the same beam not only predicts significantly more displacement in the beam, but it also reveals a maximum in the load–displacement curve. As in the case of the example of yielding in the box section discussed earlier in this chapter, such a maximum load could clearly constitute failure. In contrast to this earlier failure discussion associated with yielding, the maximum load in this case is caused by collapse of the beam's cross section in the vicinity of the applied loads, as illustrated in the cross sectional view in Fig. 5.61. Yielding in the plastic could serve to lower this collapse load, but the maximum would still exist even for an elastic material. To prevent such collapse, a designer might mold bulkhead structures into the

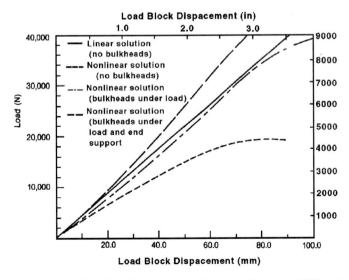

Figure 5.60 Predicted load–displacement behavior for several different box-beam geometries.

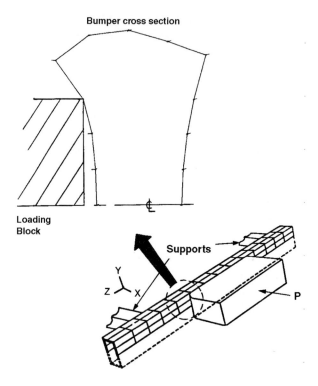

Figure 5.61 Cross-sectional collapse of box beam.

cross section of the bumper, as shown in Fig. 5.62. Such structures serve the purpose of limiting the collapse of the beam's cross-sectional geometry and thereby raise or eliminate the failure load associated with this mechanism.

If such bulkheads are provided in the central 40 cm (16 in) of the beam, the load–displacement behavior and the maximum in the curve can be significantly affected as is illustrated in Fig. 5.60. One approach to analyzing this behavior is to model the bulkheads using "gap elements" or contact elements, which are often available in commercial finite-element codes. As discussed in Chap. 3, modeling of this type is nonlinear in nature. These elements will provide no resistance until a predetermined amount of cross-sectional crushing has occurred. This crushing displacement can be defined as the clearance between front bumper face and internal structure. However, after this "gap" has been closed, the elements carry force directly from the front face of the beam to the contacted internal structure.

Modeling the bulkheads in this fashion, the beam is not only pre-

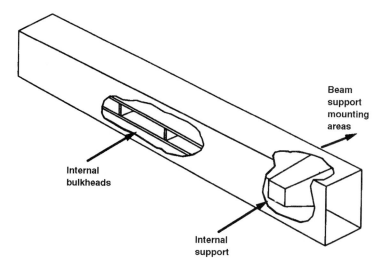

Internal
bulkheads

Beam
support
mounting
areas

Internal
support

Figure 5.62 Conceptual drawing of a box beam with internal bulkheads to prevent cross-sectional collapse.

dicted to be stiffer, but the load-limiting behavior associated with collapse in the central regions of the beam is eliminated, as can be seen in Fig. 5.60. Although cross-sectional collapse is prevented, there is still evidence of another significant structural instability. As can be seen in Fig. 5.63, the front face of the box section has buckled because of the large compressive stresses induced there under the bending load. Since this type of plate buckling is stable in nature,[70] the front face does continue to carry increased stress (below yield stress) and additional load is still possible after buckling. However, a designer might still wish to preclude such behavior. It may also be advantageous to employ bulkheads in the area of the beam supports, since locally high loads, distortion, and possible collapse are also likely in this area. If this design option is modeled using similar techniques, there is still another significant increase in the stiffness of the overall system, as is shown in Fig. 5.60.

As we saw from this example, proper understanding of design considerations unique to the use of thermoplastic materials can be essential to efficient design of thermoplastic components. Although buckling and collapse are certainly not issues normally considered in the design of a metal automotive bumper beam, they may be significant when low modulus thermoplastic materials are used. However, when issues such as the nonlinear effects of the low moduli are well understood, effective designs using these materials are possible. Many thermoplastic

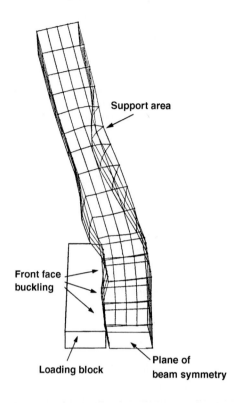

Support area

Front face buckling

Loading block

Plane of beam symmetry

Figure 5.63 Deformed shape of a box beam loaded by a rigid, centrally located block showing buckling of the front face (one half of the beam shown; symmetry assumed).

bumper beams currently include a significant amount of internal structure such as that illustrated in Fig. 5.62. Furthermore, engineering analyses of these structures have proven to be quite accurate,[69] provided the analyses are capable of accounting for large deformation nonlinearities such as buckling and collapse. We will examine some of the dynamic aspects of the impact of thermoplastic beams in Chap. 6.

Other Engineering Considerations Pertinent to Strength and Failure

There are a number of other general considerations with a bearing on strength of plastics that merit qualitative discussion. First, the processing conditions under which the component is created can have a significant effect upon the strength of the material. Excessive temperatures in the barrel of an injection-molding machine can reduce the strength and increase the component's transition temperature from ductile to brittle failure. Processing requirements vary for different plastics and must be well understood. In severe situations, a normally ductile material may fail in a brittle fashion, especially during impact loading.

In addition to processing temperature, the engineer must also consider the potential existence of *weld lines* in the molded part. A weld line is a locus of points in a molding at which two fronts of molten plastic meet during the molding process. The strength of these weld lines will generally be inferior to that of the plastic around them as measured by standard tests on molded coupons without weld lines.

Plastics are susceptible to chemical attack from a variety of agents and a common effect of this reaction is a severe embrittlement of the material. In choosing a plastic for a particular component application, the engineer should consider the chemical environment in which the component will reside. The engineer must then assess the effects that any chemicals associated with that environment will have upon the plastic. In general, amorphous thermoplastics are more susceptible to chemical attack than semicrystalline thermoplastics.

"Aging" or long-term material degradation is another chemical or material process that can occur in thermoplastics over their service lives, especially if the end-use environment includes somewhat elevated temperatures. This process also results in reduction of strength. Although these topics are more closely associated with materials science and therefore beyond the scope of this text, an engineer designing with thermoplastic materials must be aware of their existence and effects.

Closure

Once an adequate ability to predict deformation and stiffness exists, engineering analysis tools can be used by engineers to predict failure. In order to accomplish this goal, material data defining failure in terms of stress, strain, or fracture toughness must exist. In this chapter some of the most important modes of failure have been examined along with engineering criteria and capabilities available to predict such events. For some materials, such as short-glass-fiber-filled thermoplastics, failure may occur abruptly at the end of the linear-elastic range of deformation, while failure in other engineering thermoplastics occurs only after the accumulation of very large strains that are related to stresses in a very nonlinear manner.

Although linear-elastic constitutive models and simple criteria based on maximum principle stress or strain may be sufficient to predict failure in glass-filled, injection-molded thermoplastics, there are complicating issues associated with measuring realistic strength properties useful in practical applications. Preferred fiber orientation in common tensile test specimens can lead to incomplete and nonconservative representations of strength for a real component. The variation of this strength is qualitatively understood to be at least partially related to orientation of the short fibers during the injection-molding process. Although definitive theories encompassing both processing mechanics and failure criteria do not exist, there are engineering approaches to account for these effects in an approximate manner.

In contrast to linear-elastic failure, many engineering thermoplastics fail only after accumulating large strains with extremely nonlinear relationship to stress. For some of these materials, such as polycarbonate, classical plasticity theory can provide an adequate framework to predict maximum loads and permanent deformation in general components. Either of these situations may constitute failure in an engineering sense. However, failure as defined by rupture may involve behavior quite different from that normally observed in metals. Many thermoplastics, for example, exhibit propagating necks (thinner regions) in contrast to the localized necks observed in many ductile metals. This phenomenon can lead to much larger elongation and energy absorption than is evident in metals before failure. It should be emphasized that classical plasticity theory may not apply equally well for all thermoplastics that display nonlinear stress–strain behavior and significant strain to failure. Definitive categorization of which thermoplastics are adequately modeled with classical plasticity does not exist, and some care should be given to its application even though results may be very adequate in many situations.

In addition to damage as characterized by plasticity, an alternative type of damage that is observed in many thermoplastics, particularly those of the amorphous class, is crazing. Crazing is a particularly dangerous form of damage in these materials since it may lead to catastrophic fracture at deformation levels that are uncharacteristically low for ductile thermoplastics. There are several factors that have an effect on whether or not craze formation will occur. One of the most significant parameters affecting craze formation from the engineering point of view is component geometry. In general, geometries that induce stress states in the first octant of three-dimensional stress space (all three principle stresses positive) are the most likely to promote local crazing. Although the appearance of crazing does not guarantee catastrophic brittle failure in ductile thermoplastics, it does appear to be a significant contributor. The notched beam is a good example of one geometry that promotes stress states particularly susceptible to crazing and brittle failure. As discussed in this chapter, although a flat disk of polycarbonate may undergo large permanent deformation before rupturing, an appropriately notched polycarbonate beam can fail in a very brittle fashion. This phenomenon is generally referred to as notch sensitivity. Although a criterion with the same levels of acceptability and generality as the von Mises yield condition does not exist, it does seem possible to categorize problematic geometries.

In addition to geometry, temperature also plays a significant role in determining whether or not a thermoplastic will fail in a ductile or brittle fashion. In general, lower temperatures promote brittle failure. However, since this transition in failure is also stress-state and geometry dependent and since a completely confident criterion for the brittle-failure condition has not been accepted, an engineering prediction of transition temperature for a general structural component is not currently possible.

The methodology of fracture mechanics has also been shown to be applicable to thermoplastics. The temperature range over which linear-elastic fracture mechanics is valid will vary as a function of material. Unlike other engineering failure criteria, applying this methodology requires knowledge of the size of a crack in the component. As a result, this methodology is useful in assessing the ability of a damaged structure to sustain load (damage tolerance). This methodology of measuring valid fracture toughness properties useful in general engineering geometries is much more demanding than that for other strength-related properties such as yield strength.

Linear elastic fracture mechanics has limited ranges of applicability (as defined in terms of temperature, for example) for many thermoplastic materials. As a result, nonlinear fracture mechanics techniques such as the J-integral approach have been examined as a means of expanding the range of applicability. As has already been emphasized, linear elastic fracture data are much more difficult to measure accurately than most other engineering material data. In addition, to apply this data practically, some manner of defining crack sizes, such as nondestructive testing, is required. This is not commonly warranted for thermoplastic structures today. All of these inhibiting issues for linear elastic fracture mechanics are equally pertinent for nonlinear fracture methodologies. Furthermore, the test methodologies to measure true material data for a nonlinear fracture mechanics methodology are even more complex than those associated with linear-elastic fracture mechanics and also less well established. As a result, there is currently no practical, verified methodology for the use of nonlinear fracture mechanics properties of thermoplastics within the context of common part design approaches.

In addition to these material failure modes, the low modulus of most thermoplastics makes buckling instability an important structural failure mode to assess when using these materials. This is particularly important since so many structural thermoplastic components are thin walled in nature.

It will be noted that time-dependent aspects of failure, including both impact and creep phenomena, have not been discussed here. These engineering failures, as well as fatigue failure, will be examined individually in chapters to follow. Nonetheless, many of the failure phenomena introduced in this chapter including yield, rupture after large deformation, and notch sensitivity will be extremely relevant in those chapters, especially the one addressing impact events.

Finally, it is easily recognizable because of the lack of discussion that there are numerous issues of failure in thermoplastics that are not well developed with respect to the successful engineering application of failure criteria. The prediction of failure in foamed thermoplastics as well as long-glass-fiber thermoplastic sheet are two good examples. Although the failure of such materials in simple coupon form has been studied,[71–73] a consistent engineering approach for application of these data to general component geometries does not appear to have been demonstrated.

References

1. G. Ambur and G. G. Trantina, "Structural Failure Prediction with Short-Fiber Filled, Injection Molded Thermoplastic," *Proceedings of the 1988 Society of Plastics Engineers (SPE) Annual Technical Meeting*, SPE, Brookfield Center, CT, 1988, p. 1507.
2. R. Hill, *The Mathematical Theory of Plasticity*, Oxford University Press, London, 1956.
3. L. M. Kachanov, *Fundamentals of the Theory of Plasticity*, MIR, Moscow, 1974.
4. A. Mendelsohn, *Plasticity; Theory and Applications*, Macmillan, New York, 1968.
5. I. M. Ward, *Mechanical Properties of Polymers*, Wiley, Chichester, 1971, pp. 330–337.
6. A. F. Yee and L. M. Carapellucci, "The Biaxial Deformation and Yield Behavior of BPA-Polycarbonate: Effects of Anisotropic and Loading Rate," *Proceedings of the 1983 Society of Plastics Engineers (SPE) Annual Technical Meeting*, SPE, Brookfield Center, CT, 1983, pp. 370–373.
7. V. K. Stokes and H. F. Nied, "Solid Phase Sheet Forming of Thermoplastics—Part I: Mechanical Behavior of Thermoplastics to Yield," *Journal of Engineering Materials and Technology*, 108(April): 107, 1986.
8. R. E. Robertson, "On the Cold-Drawing of Plastics," *Journal of Applied Polymer Science*, (2): 443–450, 1963.
9. I. H. Müller and Jäckel, "Energie-Bilanz bei Kallierstrechung," *Kolloid Zeit*, 129(Dec.): 145–146, 1952.
10. I. Marshall and A. B. Thompson, "The Cold-Drawing of High Polymers," *Proceedings of the Royal Society of London Ser. A*, 221(Feb.): 541–557, 1954.
11. N. Brown and J. M. Ward, "Load Drop at the Upper Yield Point of a Polymer," *Journal of Polymer Science Pt. A-2*, 6(3): 607–620, 1968.
12. P. I. Vincent, "The Necking and Cold-Drawing of Rigid Plastics," *Polymer*, 1(March): 7–19, 1960.
13. H. F. Nied and V. K. Stokes, "Solid Phase Sheet Forming of Thermoplastics—Part II: Large Deformation Post-Yield Behavior of Plastics," *Journal of Engineering Materials and Technology*, 108: 113, 1986.
14. P. S. Theocaris and C. Hadjiiosiph, "Strain Analysis of Neck Formation and Propagation in Glassy Polymers," *Engineering Fracture Mechanics*, 12: 241–252, 1979.
15. D. Lee and P. C. Luken, "Material Modelling and Solid Phase Forming of Polycarbonate Sheet," *Polymer Engineering and Science*, 26: 612–619, 1986.
16. G. W. Halldin and Y. C. Lo, "The Solid-Phase Flow Behavior of Ductile Thermoplastics," *Proceedings of the 1983 Society of Plastics Engineers (SPE) Annual Technical Meeting*, SPE, Brookfield Center, CT, 1983, pp. 366–367.

17. C. G'Sell, and J. J. Jonas, "Determination of the Plastic Behavior of Solid Polymers at Constant True Strain Rate," *Journal of Materials Science,* 14: 583–591, 1979.
18. R. P. Nimmer, G. Tryson, and H. Moran, "Impact Response of a Polymeric Structure," *Proceedings of the 1984 Society of Plastics Engineers (SPE) Annual Technical Meeting,* SPE, Brookfield Center, CT, 1984, pp. 565–568.
19. G. G. Trantina and M. D. Minnichelli, "The Effect of Nonlinear Material Behavior on Snap Fit Design," *Proceedings of the 1987 Society of Plastics Engineers (SPE) Annual Technical Meeting,* Brookfield Center, CT, 1987, pp. 438–441.
20. I. M. Ward, *Mechanical Properties of Polymers,* Wiley Chichester, 1971, pp. 329–337.
21. J. W. Hutchinson and K. W. Neale, "Neck Propagation," *Journal of Mechanics and Physics of Solids,* 31 (5): 405–426, 1983.
22. R. P. Nimmer and L. Miller, "Neck Propagation in Tensile Tests," *Journal of Applied Mechanics,* 51: 759, 1984.
23. B. D. Coleman and D. C. Newman, "On the Theory of Cold Drawing and Neck Formation in Elastic Films," *Mechanics of Plastics and Plastic Composites,* Applied Mechanics Division, American Society of Mechanical Engineers, New York, 1989, Vol. 104, pp. 45–54.
24. B. S. Bagepalli, "Finite Strain Elastic-Plastic Deformation of Glassy Polymers," D. Sc. Thesis, Department of Mechanical Engineering, Massachusetts Institute of Technology, 1983.
25. B. S. Bagepalli, A. S. Argon, and D. M. Parks, "Large Elastic-Plastic Deformation of Glassy Polymers; Part I: Constitutive Modelling," MIT Program in Polymer Science and Technology Report, March 1985.
26. D. M. Parks, A. S. Argon, and B. S. Bagepalli, "Large Elastic-Plastic Deformation of Glassy Polymers; Part II: Numerical Analysis of Neck Drawing," MIT Program in Polymer Science and Technology Report, March 1985.
27. M. C. Boyce, D. F. Parks, and A. S. Argon, "Large Strain Inelastic Deformation of Glassy Polymers; Part I: Rate Dependent Constitutive Model," *Mechanics of Materials,* 7: 17, 1988.
28. S. D. Batterman and J. L. Bassani, "Yielding, Anisotropy and Deformation Processing of Polymers," *Mechanics of Plastics and Plastic Composites,* edited by V. J. Stokes, Applied Mechanics Division, American Society of Mechanical Engineers, City, New York, 1989, vol. 104, pp. 13–28.
29. A. K. Ghosh, "Plastic Flow Properties in Relation to Localized Necking in Sheets," *Mechanics of Sheet Metal Forming,* Plenum, New York, 1978, pp. 287–312.
30. A. F. Storace, R. P. Nimmer, and R. Ravenhall, "Analytical and Experimental Investigation of Bird Impact on Fan and Compressor Blading," *Journal of Aircraft,* 21 (7) 520–527, 1984.
31. R. P. Nimmer, "Analysis of the Puncture of a BPA-Polycarbonate Disc," *Polymer Engineering and Science,* 23: 155, 1983.
32. R. P. Nimmer, "An Analytic Study of Tensile and Puncture Test Behavior as a Function of Large-Strain Properties," *Polymer Engineering and Science,* 27: 263–270, 1987.
33. R. P. Nimmer, "Predicting Large Strain Deformation of Polymers," *Polymer Engineering and Science,* 27: 16–24, 1987.
34. L. M. Carapellucci, A. F. Yee, and R. P. Nimmer, "Some Problems Associated with the Puncture Testing of Plastics," *Journal of Polymer Engineering and Science,* 27: 773–780, 1987.
35. R. P. Nimmer and J. T. Woods, "An Investigation of Brittle Failure in Ductile, Notch-Sensitive Thermoplastics," *Polymer Engineering and Science,* 32 (16):1126–1137, 1992.
36. I. M. Ward, *Mechanical Properties of Polymers,* Wiley, Chichester, 1971, pp. 413–424.
37. S. Rabinowitz and P. Beardmore, "Craze Formation and Fracture in Glassy Polymers," *Critical Reviews in Macromolecular Science,* edited by E. Baer, CRC, Cleveland, OH, 1972, Vol. 4, p. 1.
38. R. P. Kambour, "A Review of Crazing and Fracture in Thermoplastics," *Journal of Polymer Science Pt. D,* 7: 1–154, 1973.
39. S. S. Sternstein and L. Ongchin, "Yield Criteria for Plastic Deformation of Glassy

High Polymers in General Stress Fields," *Polymer* Preprints, American Chemical Society, Division of Polymer Chemistry, 1969, Vol. 10, No. 2, pp. 1117–1124.

40. S. S. Sternstein and F. A. Myers, "Yielding of Glassy Polymers in the Second Quadrant of Principle Stress Space," *Journal of Macromolecular Science-Physics*, B8 (3–4): 539–571, 1973.

41. R. J. Oxborough and P. B. Bowden, *Philosophical Magazine*, 28: 547, 1973.

42. A. S. Argon, "Physical Basis of Distortional and Dilatational Plastic Flow in Glassy Polymers," *Journal of Macromolecular Science-Physics*, B8 (3–4): 573–596, 1973.

43. A. M. Guarde and V. Weiss, "Brittle Crack Initiation at the Elastic-Plastic Interface," *Metallurgical Transactions*, 3: 2811–2817, 1972.

44. N. J. Mills, "The Mechanism of Brittle Fracture in Notched Impact Tests on Polycarbonate," *Journal of Material Science*, 11: 363–375, 1976.

45. M. Ishikawa, I. Narisawa, and H. Ogawa, "Criteria for Craze Nucleation in Polycarbonate," *Journal of Polymer Science: Polymer Physics Edition*, 15: 1791–1804, 1976.

46. I. M. Narisawa and H. Ogawa, "Notch Brittleness of Ductile Glassy Polymers Under Plane Strain," *Journal of Material Science*, 15: 363–375, 1980.

47. M. Ishikawa, H. Ogawa, and I. Narisawa, "Brittle Fracture in Glassy Polymers," *Journal of Macromolecular Science-Physics* B19 (3): 421–443, 1981.

48. M. Ishikawa and I. Narisawa, "Fracture of Notched Polycarbonate Under Hydrostatic Pressure," *Journal of Materials Science*, 18: 1947–1957, 1983.

49. M. Ishikawa and I. Narisawa, "The Effect of Heat Treatment on Plane Strain Fracture of Glassy Polymers," *Journal of Materials Science*, 18: 2826–2834, 1983.

50. R. P. Kambour, M. A. Vallance, E. A. Farraye, and I. A. Grimaldi, "Heterogeneous Nucleation of Crazes Below Notches in Glassy Polymers," *Journal of Material Science*, 21: 2435–2440, 1986.

51. R. Hill, *The Mathematical Theory of Plasticity*, Oxford University Press, London, 1950, pp. 128–136 and 245–252.

52. J. R. Griffiths and D. R. Owen, "An Elastic Plastic Stress Analysis for a Notched Bar in Plane Strain Bending," *Journal of Mechanics and Physics of Solids*, 19: 419–431, 1971.

53. J. T. Woods, and H. G. deLorenzi, "An Assessment of Crazing Criteria for Polyetherimide in Three-Dimensional Stress Space," *Proceedings of the 1992 Society of Plastics Engineers (SPE) Annual Technical Meeting*, SPE, Brookfield Center, CT, 1992, pp. 1724–1727.

54. I. M. Ward, *Mechanical Properties of Solid Polymers*, Wiley, 1983, pp. 424–432.

55. P. I. Vincent, *Polymer*, 1: 427, 1960.

56. *1993 Annual Book of ASTM Standards*, Section 8, Volume 08.03, "Plastics—(III)," American Society for Testing of Materials, Philadelphia, 1993, pp. 314–322.

57. H. Tada, P. C. Paris, and G. R. Irwin, *The Stress Analysis of Cracks Handbook*, Del Research Corp., Hellertown, PA, 1973.

58. L. N. Gifford and P. D. Hilton, "Stress Intensity Factors by Enriched Elements," *Engineering Fracture Mechanics*, 10: 485–496, 1978.

59. S. E. Benzely, "Representation of Singularities with Isoparametric Finite Elements," *International Journal for Numerical Methods in Engineering*, 8: 537–545, 1974.

60. R. D. Henshell and K. G. Shaw, *International Journal of Numerical Methods in Engineering*, 9: 495–507, 1975.

61. R. S. Barsoum, *International Journal of Numerical Methods in Engineering*, 10: 25–37, 1976.

62. J. M. Bloom, *International Journal of Fracture*, 11: 705–707, 1975.

63. C. F. Shih, H. G. deLorenzi, and M. D. German, *International Journal of Fracture*, 12: 647–651, 1976.

64. W. F. Brown and J. E. Scrawley, *Plane Strain Crack Toughness Testing of High Strength Metallic Materials*, ASTM STP 410, ASTM, Philadelphia, 1966.

65. J. G. Williams, *Fracture Mechanics of Polymers*, Ellis Horwood, Chichester, 1984.

66. A. J. Kinloch and R. J. Young, *Fracture Behavior of Polymers*, Applied Science, London, 1983.

67. R. P. Nimmer, K. Weiss, J. McGuire, M. Takemori, and J. Morelli, "Engineering Application of Linear Elastic Fracture Mechanics for a Ductile Polymer Part," *Proceed-

ings of 1988 Society of Plastics Engineeers (SPE) Annual Technical Meeting, SPE, Brookfield Center, CT, 1988, pp. 1503–1506.

68. *1993 Annual Book of ASTM Standards,* Section 3, Volume 03.01, "Metals—Mechanical Testing," American Society for Testing and Materials, Philadelphia, 1993, pp. 509–539.

69. R. P. Nimmer, O. A. Bailey, and T. W. Paro, "Analysis Techniques for the Design of Thermoplastic Bumpers," Society of Automotive Engineers (SAE) Tech. Paper Series, Paper No. 870107, SAE, Warrendale, PA, 1987.

70. D. Brush and B. O. Almroth, *Buckling of Bars, Plates and Shells,* McGraw-Hill, New York, 1975, p. 89.

71. R. P. Nimmer, V. K. Stokes, and D. A. Ysseldyke, "Mechanical Properties of Rigid Thermoplastic Foams—Part II: Stiffness and Strength Data for Modified Polyphenylene Oxide Foams," *Polymer Engineering and Science,* 28: 1501–1508, 1988.

72. V. K. Stokes, "Random Glass Mat Reinforced Thermoplastic Composites: Part IV, Characterization of the Tensile Strength," *Polymer Composites,* 11: 354–367, 1990.

73. W. C. Bushko and V. K. Stokes, "Strength of Glass Mat Reinforced Thermoplastic Composites—A Statistical Approach," *Proceedings of the 1992 Society of Plastics Engineers (SPE) Annual Technical Meeting,* SPE, Brookfield Center, CT, 1992, pp. 783–787.

[faded, illegible bibliographic entries]

6

Designing for Impact
with Plastic Materials

Impact resistance is quite often an engineering quality that plays a significant role in the design process of a plastic component. In many applications it may actually be the primary criterion for design. Automotive bumpers must fulfill requirements for surviving collisions while simultaneously protecting the surrounding automotive subcomponents. Many of these bumpers are made using plastic materials. Instrument panels inside the car are also often made of a plastic material. Design considerations in this case often focus on minimizing occupant injury during impact. Airplane canopies, electronic equipment chassis, industrial storage pallets, and football helmets are just a few of the many examples of plastic parts that must be impact resistant to be successful.

In this chapter we will be dealing with material and engineering issues associated with plastic components subjected to impact events. As a starting point for this discussion, we must consider what characterizes or defines an impact event. With a working definition of an impact event at hand, it will be seen that the rate of loading is significant to these events in two ways. First, if the rate of loading is fast enough, inertial (mass) effects of the structures may be significant. Second, since the mechanical properties of plastics are often rate sensitive, it becomes important to quantify how rate affects these properties. Naturally any effort to predict the performance of an engineering component must be fundamentally connected to measurable material performance. There are a number of tests that customarily are associated with the impact response of plastic materials, including the standard tensile test, the Gardener impact test, and the notched Izod or Charpy impact test. Some of these aforementioned tests actually measure true material properties that can be applied in engineering analyses, while others provide much more qualitative information. These tests and the various issues associated with them will also be discussed in this chapter. Finally, a number of examples will be presented that represent practical experience in applying these engineering analysis techniques and material properties to actual impact events. Results of these predictions will be compared to test results.

Definition of Impact Events

Generally, an impact problem involves at least two flexible objects, each with its own geometry, stiffness, and mass distribution, traveling on a collision course with velocities V_1 and V_2 as shown in Fig. 6.1. Extensive discussion of the engineering treatment of impact events can be found in Refs. 1 and 2. There are a number of issues characteristically different about this event in comparison to other common engineering situations. First of all, when two objects collide, each will be subject to forces that are suddenly applied over a short period of time and then removed in an equally rapid fashion. The time of loading, described graphically in Fig. 6.2, distinguishes an impact load from the more slowly applied static load, also shown in the figure. Differentiation of what constitutes a rapid or slow loading rate can be based on different norms, and we will discuss the meaning of a "short" period of time more quantitatively later.

The high rates of loading that are normally associated with impact events lead to two potential consequences that must be considered with respect to carrying out engineering analyses. First, it is often neces-

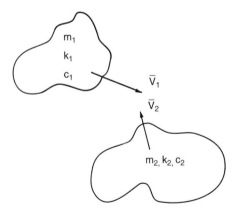

Figure 6.1 General two-body impact event (mass, m; stiffness, k; damping, c).

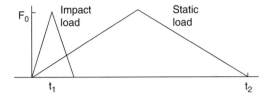

Figure 6.2   Qualitative comparison of impact load and static load.

sary to include the effects of mass in the analysis, and second, the rate sensitivity of the governing material properties may also be a consideration. Both of these issues will be discussed generally as well as specifically in relation to their effect for plastic materials.

There is another significant difference between the impact problem illustrated in Fig. 6.1 and other common engineering problems. Although most engineering design and analysis issues are defined in terms of the performance of a structure with respect to a well-specified load condition, the load history associated with an impact event may have to be derived as part of the solution of a general impact problem. If the geometry or material of either body is altered, especially in the area of impact, the loads associated with that impact may also be significantly altered. As a consequence, impact performance criteria are usually defined in terms of impact velocities, masses, distances dropped, and energy absorption requirements. This form of "load" specification leads to a conceptually different process of formulating the analysis approach for a design. An analyst solving problems of this class must be aware of this fundamental interaction and apply modeling and analysis techniques that account for it.

Impact resistance can be considered to be the relative resistance of a component to failure due to stresses applied at high rates. Failure of a plastic component under stress can take many forms. There are situations where excessive elastic deformation will constitute failure, such as seen in a plastic automotive bumper. A bumper system is required to absorb specified levels of energy while simultaneously protecting the rest of the automobile from damage. If the plastic bumper withstands an impact without damage, but it undergoes such a large displacement that it contacts and dents the automobile's sheet metal, then it has failed in its function. In other applications, the criteria defining impact failure may be linked directly to damage, such as yielding, crazing, and fracture, all of which were discussed in Chap. 5. If a plastic panel is to be used in an exterior automobile body application, the ability to withstand an impact without denting may be an appropriate measure of failure. In other situations, visible damage may not constitute failure as long as the plastic component has not been punctured or penetrated by an impactor. Plastic containers or impact-resistant plastic window glazing might be subject to this type of failure criteria. Perhaps the most catastrophic failure associated with impact occurs when a plastic component shatters or at least fractures in a brittle manner. This event is usually intolerable and every engineering precaution must be taken to avoid it.

Impact Response: Transient Dynamic Effects

One of the essential characteristics for understanding impact events is the role of mass (inertia) in a component's response. There are several aspects to consider and an appropriate place to start is with the impact response of a simple, idealized, discrete system, such as the one shown in Fig. 6.3. The dynamic response of such a system is discussed in any elementary text on dynamics, such as Ref. 3. The simple system shown in Fig. 6.3 is composed of three elements: a mass m, a spring with stiffness k, and a dashpot with a coefficient of viscous damping c. The dynamic equation of equilibrium for such a system is

$$m\ddot{x} + c\dot{x} + kx = F(t) \qquad (6.1a)$$

$$x(0) = x_0 \qquad (6.1b)$$

$$\dot{x}(0) = v_0 \qquad (6.1c)$$

where each dot over a variable represents differentiation with respect to time.

The solution to Eq. (6.1) is usually written in terms of two parameters, ω and ζ, where ω is the system's natural frequency defined as

$$\omega = \sqrt{k/m} \qquad (6.2)$$

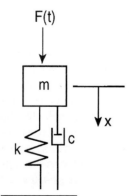

Figure 6.3 Idealized one-degree-of-freedom dynamic system.

and ζ is the viscous damping factor defined as

$$\zeta = c/2m\omega \qquad (6.3)$$

The forcing function $F(t)$ in Eq. (6.1a) is completely general in nature. However, in the case of an impact load, the time during which the load is applied is generally short in duration. In order to judge whether the duration of a load has been "short," there must be some standard of time with which the loading time can be compared. One logical measure of time to use for this comparison is the natural period of oscillation for the unforced response of the undamped dynamic system.

If there is no load applied to the system in Fig. 6.3, but the mass is moved away from its equilibrium position by a distance x_0 and released, then the solution to the differential equation defined by Eq. (6.1) is

$$x = \frac{x_0\, e^{-\zeta\omega t}}{\sqrt{1-\zeta^2}} \cos\left[\sqrt{1-\zeta^2}\,\omega t - \tan^{-1}\!\left(\frac{\zeta}{\sqrt{1-\zeta^2}}\right)\right] \qquad (6.4)$$

In such a case the response of the mass is a decaying sinusoidal function of time that is shown in Fig. 6.4 for a damping factor ζ of 0.1. If

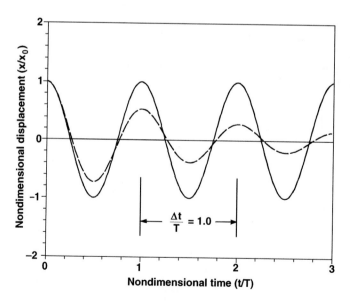

Figure 6.4 Damped and undamped response of a single-degree-of-freedom system initially displaced from its equilibrium position.

there is no damping in the system, the dynamic response does not decay with time, as is also shown in Fig. 6.4. For any small damping factor ($\zeta \ll 1.0$), the natural period of the sinusoidal portion of the response function is only negligibly different from the undamped response, as shown in the figure. This natural period can be shown to be

$$T = 2\pi/\omega = 2\pi\sqrt{m/k} \qquad (6.5)$$

The regular oscillation of a free dynamic system shown in Fig. 6.4 and characterized by the natural period defined in Eq. (6.5) establishes a natural "clock" for the system with the time period T as the basic unit of time. One method of judging whether an event such as a load application is fast or slow is to compare the loading time to the natural period of the dynamic system.

Using the simple system shown in Fig. 6.3, we can more quantitatively define the meaning of a "short" load time as well as its consequences. Assume that the load $F(t)$ applied in Fig. 6.3 is defined as

$$F(t) = \begin{cases} 0 & t \le 0 \\ F \sin \pi\dfrac{t}{T_L} & 0 \le t \le T_L \\ 0 & t \ge T_L \end{cases} \qquad (6.6)$$

This load is shown graphically in Fig. 6.5. The general displacement response x of the single-degree-of-freedom system to this simple load can be expressed as

$$x(t) = \int_0^t F(\xi)\frac{e^{-\zeta\omega(t-\xi)}}{m\omega\sqrt{1-\zeta^2}} \sin[\sqrt{1-\zeta^2}\,\omega(t-\xi)]\,d\xi \qquad (6.7)$$

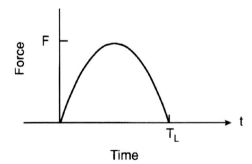

Figure 6.5 Half-sinusoid impact load.

First, consider the response of this system for a variety of impact load duration times T_L in order to understand quantitatively what defines a "short" load time. For this comparison, we will assume that the damping coefficient ζ is 0.05, the natural frequency ω is 2π, and the maximum applied load F is $(2\pi)^2$. Using the definition given in Eq. (6.5), the natural period for this system is

$$T = 2\pi/\omega = 1.0 \tag{6.8}$$

For such a system and maximum applied force, the system displacement x is illustrated graphically in Fig. 6.6 for loading times T_L of 5.0, 2.0, 1.0, and 0.5. The ordinate of Fig. 6.6 is the dynamic displacement nondimensionalized by the displacement under a static load of equal magnitude F. If the load duration is $5 \times$ (the natural period of the system), the response shown in Fig. 6.6 is nearly static in nature. That is to say, the displacement at any time is quite accurately approximated by the force defined in Fig. 6.5 divided by the system stiffness k. There is some visible oscillation with a frequency equal to the system natural frequency during the application of load and some small free vibration

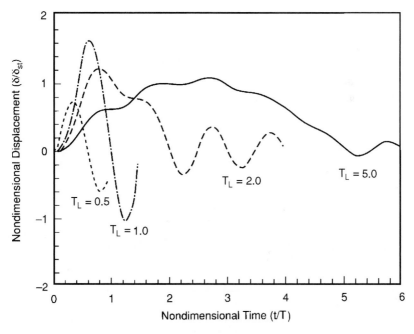

Figure 6.6 Impact response of an idealized dynamic system as a function of the duration of load.

after the dynamic load is removed, but both are negligible. As the load duration T_L is decreased while the maximum force is held constant, there are noticeable differences in the dynamic response. First, there is noticeable free vibration of the system after the loading is terminated. Second, as the load duration is shortened and approaches the natural period of the system, the displacement during the load is noticeably different than the equivalent static displacement. As the load durations approach the system natural period, the maximum dynamic displacement becomes significantly larger than the equivalent static displacement. As the time duration is further decreased to values well under the first natural period, the maximum dynamic response becomes less than the equivalent static response. Considering this comparison in Fig. 6.6, an impact load could be described as a load with a time of application on the same order of or smaller than the period of the dynamic system. For load times considerably larger than the system natural period, the mass of the system may be ignored and the forces and displacements calculated using static equations. However, for shorter loading times, the inertial effects must be considered because the system dynamic response may be substantially larger than a static approximation.

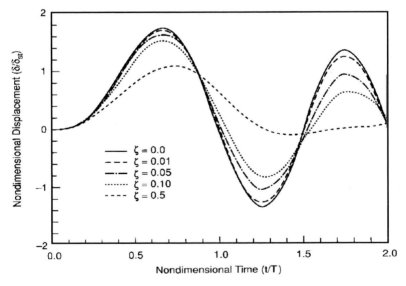

Figure 6.7 Impact response of an idealized dynamic system as a function of damping.

Damping effects

In the previous discussion examining the effect of the load application time for an impact event, the nondimensional damping coefficient ζ was assumed to be 0.05. Although the stiffness k and mass m of dynamic systems are usually well defined, the damping coefficient is usually more difficult to quantify. Figure 6.7 illustrates the effect of the damping coefficient upon the dynamic response of the system. As in Fig. 6.6, the displacement x is nondimensionalized by the static displacement under force F, and the time is nondimensionalized by the natural period of response. For this comparison, the load time T_L is assumed to be equal to the natural period of the system. The system response to this load is illustrated in Fig. 6.7 for five values of the damping factor ζ, ranging from 0 to 0.5. For most structural systems, ζ will have a value between 0.0 and 0.05. In this range, the maximum displacement of the system varies by only about 5%. The largest response associated with a damping coefficient of 0.0. As a result from the standpoint of conservative analysis, assuming that the damping is zero will be a reasonable first approximation for predicting the initial peak response of the system to an impact load.

Continuous systems and wave propagation

Although the spring–mass–dashpot system allows several significant aspects of dynamic response to be highlighted, it cannot properly model several additional issues associated with dynamic loading events. The significant missing characteristic is the distributed nature of the stiffness and mass. In order to illustrate the effects that distributed mass and stiffness can play in dynamic response, consider the simple bar shown in Fig. 6.8. The governing differential equation of equilibrium for the dynamic response of this bar is:

$$E\frac{\partial^2 u}{\partial x^2} = \rho\frac{\partial^2 u}{\partial t^2} \tag{6.9}$$

where E is Young's modulus, ρ is the mass density, x is the linear coordinate defining a position along the length of the bar, and t is time. In addition to the equation of equilibrium, a set of boundary conditions and initial conditions are necessary to define the problem completely. If the bar is fixed at one end and loaded with a time-varying stress at the other end, then the boundary conditions are

$$u(0,t) = 0 \tag{6.10}$$

$$\frac{\partial u}{\partial x}(L,t) = \frac{\sigma_0(t)}{E} \tag{6.11}$$

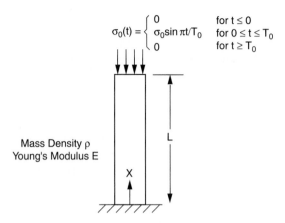

Figure 6.8 One-dimensional elastic bar.

Furthermore, if the bar is initially at rest, then the initial conditions are

$$u(x,0) = 0 \tag{6.12}$$

$$\frac{\partial u}{\partial t}(x,0) = 0 \tag{6.13}$$

For the present example, consider a half-sinusoidal stress pulse defined in Fig. 6.8. This stress will be applied at the $x = L$ coordinate position of the bar. It is applied at $t = 0$, reaches a maximum, and returns to zero again after a time duration of T_0.

The problem defined by Eqs. (6.10) to (6.13) can be solved by standard methods of separation of variables and superposition of the eigenfunctions. Using this approach, the dynamic displacement history of the bar can be expressed as

$$u(x,t) = u_p(x,t) + \sum_{n=1,3,5,\ldots}^{\infty} C_n \sin \omega_n t \, \sin(k_n x) \tag{6.14}$$

where

$$u_p(x,t) = \begin{cases} 0 & t \leq 0 \\ \dfrac{\sigma_0}{E} \dfrac{\left(\dfrac{c_0 T_0}{\pi}\right)}{\cos\left(\dfrac{\pi L}{c_0 T_0}\right)} \sin \dfrac{\pi x}{c_0 T_0} \sin \dfrac{\pi t}{T_0} & 0 \leq t \leq T_0 \\ 0 & t \geq T_0 \end{cases} \tag{6.15}$$

For $0 \le t \le I_0$ it can be shown that

$$C_n = \frac{-4\sigma_0 L}{E(n\pi)^2 \cos\left(\dfrac{\pi L}{cT_0}\right)} \left[\frac{\sin\dfrac{n\pi}{2}\left(1 - \dfrac{2L}{ncT_0}\right)}{\left(1 - \dfrac{2L}{ncT_0}\right)} - \frac{\sin n\dfrac{\pi}{2}\left(1 + \dfrac{2L}{ncT_0}\right)}{\left(1 + \dfrac{2L}{ncT_0}\right)} \right] \tag{6.16}$$

$$c_0 = \sqrt{\frac{E}{\rho}} \tag{6.17}$$

$$\omega_n = \frac{n\pi c_0}{2L} \qquad n = 1,3,5,\dots \tag{6.18}$$

$$k_n = \frac{n\pi}{2L} \qquad n = 1,3,5,\dots \tag{6.19}$$

The distributed mass and stiffness properties of this dynamic system lead to several differences in comparison to the response of the lumped mass and spring system previously considered. First of all, the bar with distributed mass has an infinite number of natural frequencies of response defined in Eq. (6.18). The spring–mass system has only one. A second significant new concept associated with the distributed mass system is the idea of wave propagation and wave speed. The reader is referred to Refs. 4 and 5 for thorough discussions of this subject. Mathematically, these concepts are closely associated with the fact that any function f of the variable ξ is a solution to the differential Eq. (6.9), provided

$$\xi = x \pm ct \tag{6.20}$$

where

$$c = \sqrt{\frac{E}{\rho}} \tag{6.21}$$

since substitution of $f(\xi)$ into Eq. (6.9) leads to the identity

$$E\frac{\partial^2 f}{\partial \xi^2} = c^2 \rho \frac{\partial^2 f}{\partial \xi^2} = \frac{E}{\rho}(\rho)\frac{\partial^2 f}{\partial \xi^2} \tag{6.22}$$

Physically, this means that if the end of a semi-infinite bar is excited such that it undergoes a displacement

$$u(0,t) = f(t) \tag{6.23}$$

Then, the displacement of the bar at some other spatial location x_1 is given by

$$u(x, t) = f(x_1 - ct) \tag{6.24}$$

The form of the excitation as a function of time at location x_1 will be the same as the form of the excitation at the bar end. However, the arrival of that excitation at x_1 will be delayed by a time

$$\Delta t = \frac{x_1}{c} \tag{6.25}$$

Consideration of Eq. (6.25) leads to an understanding of why the parameter c is understood to be the speed of the wave associated with Eq. (6.9).

Now let us consider a simple example of the dynamic response of a bar and examine the similarities and differences it has with respect to the simple spring–mass system previously discussed. Figure 6.9a to 6.9d describes the stress responses at the fixed end of the simple bar when a dynamically applied forcing function is applied at the end defined as $x = L$. In each of the four cases considered, the maximum applied stress is the same. However, the time duration of the applied load is successively shortened in each example just as it was in the previous

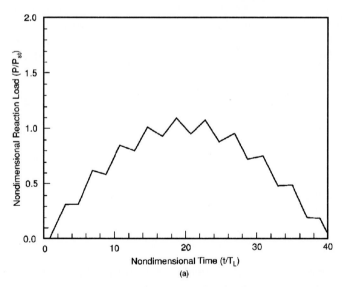

(a)

Figure 6.9 Dynamic stress response at the fixed bar end due to a half-sinusoid load at the free end; (a) load duration T_0, 10 times the first bar period T_1; (b) $T_0 = 2T_1$; (c) $T_0 = T_1$; (d) $T_0 = \frac{3}{4}T_1$.

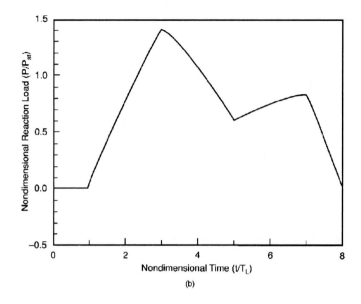

(b)

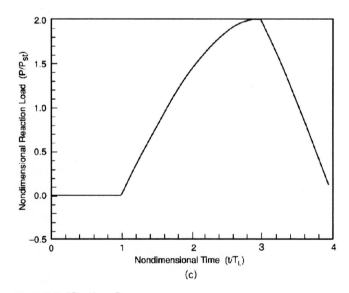

(c)

Figure 6.9 (*Continued*)

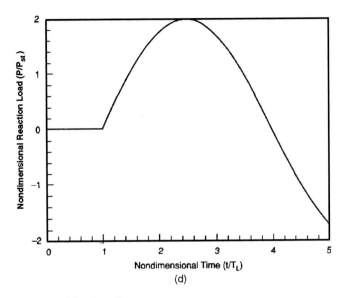

Figure 6.9 (*Continued*)

spring–mass problem. The abscissa in each case is a nondimensional time variable defined as

$$\bar{t} = \frac{tc}{L} \tag{6.26}$$

The nondimensionalizing factor in Eq. (6.26), L/c, is the time it would take a disturbance to travel from one end of the bar to the other. The ordinate is the stress response at the $x = 0$ location nondimensionalized by the maximum stress applied at the opposite end.

In the first case the load applied is imposed over a time period T_0 equal to 10 times the first natural period of the bar. This time duration is equivalent to 40 times the travel time required for a disturbance to travel from one end of the bar to the other. As can be seen from Fig. 6.9a, the maximum stress response at the opposite end of the bar is only slightly larger than the imposed stress. There is no response visible at the $x = 0$ location until after a nondimensional time of 1—the time required for a wave to travel the length of the bar. For a load applied over this length of time or longer, the response is quite accurately approximated as static in nature. The inertial effects of the bar are negligible. If the duration of the impact load application is decreased to a time equal to twice the first natural period of the bar, then the response at $x = 0$ is shown in Fig. 6.9b. As was true in the previous example, there is no response at $x = 0$ until after the time required for a disturbance to propagate the length of the bar. In addition, the response of the bar at this location is now larger than the applied load at the other end. This behavior is much like that observed in the discrete spring–mass system previously considered. If the time of load application is still further reduced to the same size as the first natural period of the bar, the magnitude of the stress response at the $x = 0$ location in the bar reaches a level twice that of the applied stress as shown in Fig. 6.9c. Additional reduction in the load application time does not result in any additional load magnification of the response at the $x = 0$ location. For shorter load times, the stress magnification factor remains at 2, as illustrated by the response of the bar to a load of duration three-quarters of the first bar period. For load durations this small and without any damping present in the system, the initial pulse of stress applied at the $x = L$ location simply propagates back and forth along the length of the bar. The magnification factor of 2 in the stress response at the $x = 0$ location is a result of the wave reflection phenomena at the fixed bar end. It should be noted that the stress magnification at $x = 0$ remains at twice the applied load for load durations less than the first natural frequency of the bar and is quite different from that observed in the discrete spring–mass model. In that case, the dynamic response of the system decreased and eventually became less than the applied load. This difference can be traced to the distributed mass property of the bar system that gives rise to wave propagation effects that are not part of the character of the discrete mass system.

Load definition in impact events

In all of the examples that we have discussed so far with respect to the dynamic response of engineering systems it has been assumed that the applied load has been well defined in terms of both magnitude and duration. As emphasized earlier, this is not generally the case in most impact events. In most cases the impact event is defined in terms of velocities, masses, or kinetic energies. The loads must usually be calculated as part of the solution process.

As a means of further exploring the analysis of impact events, consider an idealized example of an impact event shown in Fig. 6.10. There are two bodies associated with this impact, each described by a mass and a spring. Body 1 is attached to the ground with its spring and is at rest. Body 2 is moving with a velocity V toward body 1. It will be assumed that the response of body 1 is of primary interest. If the impact force imposed on body 1 is defined, then its response can be calculated in the same fashion as considered previously for the system in Fig. 6.3. However, in this case, the applied force is not explicitly defined. Its definition requires knowledge of the response of both bodies.

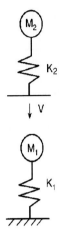

Figure 6.10 Idealized two-body impact event.

During the time the two bodies are in contact, the equations of equilibrium and the initial conditions are described as

$$M_1\ddot{x}_1 + (K_1 + K_2)x_1 - K_2\,x_2 = 0 \qquad (6.27a)$$

$$M_2\,\ddot{x}_2 + K_2\,x_2 - K_2\,x_1 \equiv 0 \qquad (6.27b)$$

$$x_1(0) = x_2(0) = 0 \qquad (6.27c)$$

$$\dot{x}_1(0) = 0 \qquad (6.27d)$$

$$\dot{x}_2(0) = V \qquad (6.27e)$$

The force exerted on body 1 can be expressed as

$$F_{\text{imp}} = K_2(x_2 - x_1) \qquad (6.28)$$

There are several important concepts highlighted by this simple problem. First, both the magnitude and the time history of the impact force are not only dependent upon the kinetic energy of the impactor (body 2), but also the stiffness of body 2 as well as the mass and stiffness of body 1. For example, if the mass and stiffness of the two bodies are respectively equal and have values of 1 unit of mass and 1 unit of stiffness, then the impact force history will have the form shown in Fig. 6.11. However, if the stiffness of body 2 is reduced to $\frac{1}{2}$ unit, then the impact force time history will have a significantly different form, also shown in Fig. 6.11. Similarly, if either the mass or the stiffness of body 1 is changed, the impact force is also altered. In the most general of situations, a transient dynamic analysis of the impact event would be necessary with both the impactor and the target bodies modeled.

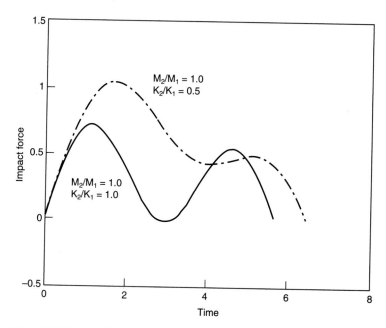

Figure 6.11 Impact force versus time histories for different stiffness ratios of a two-body system.

Some impact problems can be considered in a simpler fashion. If the mass and the stiffness of the impactor are large enough in comparison to the respective mass and stiffness of the target, then an energy balance can be used to approximate the maximum response of the target. With this approach, it is assumed that the two masses reach zero velocity at approximately the same time and at that point, the original kinetic energy of the impactor is completely converted to strain energy stored in the target. If this approximation is applied to the two-body problem just considered, then the energy balance can be written as

$$\tfrac{1}{2}\,M_2 V^2 = \tfrac{1}{2}\,K_1 x^2_{1\,max} \tag{6.29}$$

The maximum displacement of the target can then be expressed as

$$x_{1\,max} = \sqrt{\frac{M_2}{K_1}}\,V \tag{6.30}$$

As can be seen from Fig. 6.12, this approximation is very accurate even when the masses and stiffnesses of the impactor and target are equal. However, for the situations where the target mass M_1 is 5 times the impactor mass, or the impactor stiffness is $\tfrac{1}{5}$ of the target stiff-

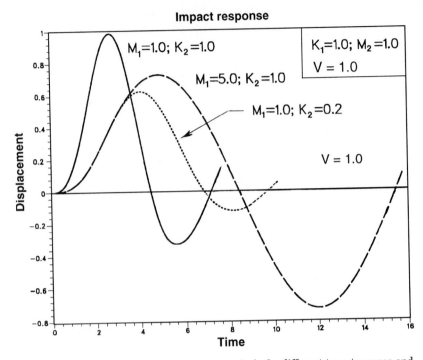

Figure 6.12 Displacement response of the target body for different target masses and impactor stiffnesses.

ness, which are also illustrated in Fig. 6.12, there are considerable errors in using this approximation.

For the case where $M_2/M_1 \gg 1.0$ and $K_2/K_1 \gg 1.0$, it can be shown that the time duration over which the impact force is applied is approximately

$$T_1 = \pi \sqrt{\frac{M_2}{K_1}} \tag{6.31}$$

Since this loading time is much longer than the natural period of the target $(T_1 = 2\pi\sqrt{M_1/K_1})$, the impact force applied to the small target mass is quasistatic in nature as pointed out previously in this chapter. As a result, a static load could be applied to the target structure to approximate the response of the target. Even if significant nonlinearity in the target occurs, the maximum deformation of the target can be identified as the point where the integrated work exerted by the force on the target is equal to the original impact energy of the projectile.

This approximation is particularly important for plastic structures since they are generally light and flexible in comparison to other structures. In many realistic engineering applications, the mass of a thermoplastic component is negligible in comparison to that of accompanying metallic subcomponents or impactors. In addition, the modulus of elasticity for common thermoplastics varies from values as low as 28 MPa (4 ksi) for some elastomers to values as high as 10.3 GPa (1500 ksi) for glass-filled plastics. Most unfilled engineering thermoplastics have moduli in the 1.4 to 3.5 GPa (200 to 500 ksi) range. All of these values are very low relative to other engineering materials. Steel and aluminum, for instance, have moduli of 207 GPa (30×10^6 psi) and 69 GPa (10×10^6 psi), respectively—one to two orders of magnitude stiffer than most thermoplastics. Furthermore, most plastic structures are thin in nature, and as a result quite flexible from a structural standpoint. As a result, it is often possible to assume that a metal member of a component or impactor is rigid in comparison to a plastic component. Examples where these assumptions are appropriate include steel balls or pendulum masses impacting plastic components like computer housings or automobile bumpers. Another similar example would be the drop testing of electronic components with plastic housings and significant, internal metallic masses. In many of these examples, the previously mentioned approximations associated with neglecting the mass of the plastic and assuming the metallic components to be rigid in comparison to the plastic may significantly simplify the necessary engineering calculations versus a complete two-body dynamic analysis.

A specific example of this type of solution strategy is worthwhile since it provides an opportunity to illustrate a significant advantage of plastics in comparison to metals with respect to impact dent resistance. If a steel ball is dropped from a height h onto a beam as shown in Fig. 6.13, the beam will bend as the ball is gradually brought to rest.

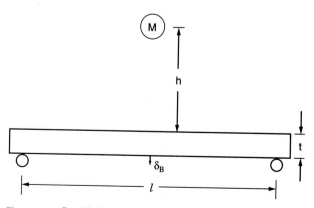

Figure 6.13 Steel ball dropped onto a beam.

With reference to our earlier discussion about two-body impacts, let us assume that the beam is much more flexible than the steel ball and the mass of the steel ball is larger than the mass of the beam. As was previously illustrated, under these conditions, this impact problem can be analyzed using an energy balance. Since the mass of the beam is small in comparison to the impacting ball, the kinetic energy of the beam during the impact can be approximated as negligible. In addition, the relative stiffness of the steel ball allows the assumption of negligible strain energy storage in the ball. As a result, the original potential energy of the ball due to its position above the beam is assumed to convert into internal strain energy of the beam when the ball is finally brought to rest. During the impact, a force P is developed between the ball and the beam that can be idealized as a concentrated load at the center of the beam. Applying the static relationship between force and deflection for a centrally loaded, simply supported beam leads to

$$\delta_B = \frac{Pl^3}{48EI} \tag{6.32}$$

where l is the length of the beam, E is the Young's modulus of the material, and I is the second moment of inertia of the beam's cross section. For this discussion, it will be assumed that the beam has a simple rectangular cross section of width b and thickness t as shown in the figure. As the beam deflects during the impact, strain energy is accumulated in the beam, and its maximum value when the ball is brought to rest can be expressed as

$$U_B = \frac{1}{2}\left(\frac{48EI}{l^3}\right)(\delta_B)^2 = \frac{1}{2}\left(\frac{l^3}{48EI}\right)P^2 \tag{6.33}$$

When the impacting ball has been brought to rest, the strain energy in the beam must be equal to the change in the potential energy of the ball from its original position to its final resting position, and the equation expressing that energy balance takes the form

$$M(h + \delta_B) = \frac{1}{2}\left(\frac{48El}{l^3}\right)(\delta_B)^2 \tag{6.34}$$

Larger strain energy storage capability means that the beam can withstand impacts from balls of larger mass dropped from greater distances.

Of interest in this discussion is the capacity of a ductile plastic beam to store energy in comparison to a ductile metal such as steel, for example. Damage in this discussion will be assumed to be "denting" due to onset of plastic deformation of the beam material. For the beam in Fig. 6.13, the maximum stress in the beam is at the center and can be expressed as

$$\sigma = \frac{6M}{bt^2} = \frac{3Pl}{2bt^2} \qquad (6.35)$$

Using Eq. (6.35), the load P_y at which the beam will first yield can be expressed as

$$P_y = \tfrac{2}{3}\frac{\sigma_y bt^2}{l} \qquad (6.36)$$

Furthermore, the strain energy stored by the beam before any plastic deformation is incurred can be described using Eqs. (6.33) and (6.35) as

$$U_y = \frac{1}{2}\left(\frac{l^3}{48E_I}\right)\left(\frac{2\sigma_y bt^2}{3}\frac{}{l}\right)^2 = \frac{1}{18}\left(\frac{\sigma_y^2}{E}\right)(btl) \qquad (6.37)$$

This expression for the elastic strain energy that can be stored in the beam before permanent deformation is dependent upon the geometry of the beam (b,L,t) and the properties of the material used in the beam (σ_y,E). Let us examine the amounts of energy that can be stored in such a beam before yield depending upon whether it is made of a ductile steel or a ductile plastic. For this comparison, it will be assumed that the steel has a Young's modulus of 210 GPa (30×10^6 psi) and a yield stress of 310 MPa (45×10^3 psi), while the plastic has a Young's modulus of 2.10 GPa (300×10^3 psi) and a yield stress of 62 MPa (9×10^3 psi).

One way to make the comparison is to assume that the geometry of the beam is fixed to be the same by other constraints regardless of the material. Under this assumption, the strain energy that can be stored without permanent deformation will scale as (σ_y^2/E). Using the material properties for steel and plastic given above, a plastic beam with the same geometry as a steel beam will be able to store 4 times the energy of the steel beam. In terms of the problem of the dropped mass, that would mean that a given mass could be dropped from 4 times the height that is required to initiate yielding for a steel beam. Alternatively, if the height is held constant, the mass that is required to initiate yielding in a plastic beam is 4 times the mass required to initiate yielding in the steel beam.

Of course, if the geometry of the beam is held constant, the stiffness of the plastic beam will be significantly less than that of the steel beam. Since stiffness is often a major factor in a part, an alternative approach to comparing the amount of energy that can be stored without yielding is on the basis of beams of equal stiffness. The stiffness of a centrally loaded, simply supported beam can be expressed as

$$K_B = \frac{P}{\delta} = \frac{48EI}{l^3} \qquad (6.38)$$

If the length and width of the beam are constrained to be the same for both the plastic and the steel beams, then, for equal stiffness, the plastic beam thickness t_p must be

$$t_p = t_s \sqrt[3]{\frac{E_s}{E_p}} \qquad (6.39)$$

where t_s is the steel beam thickness, E_s is the Young's modulus of steel, and E_p is the Young's modulus of plastic.

Using Eqs. (6.37) and (6.39), it can be seen that under the condition of equal stiffness, the strain energy that can be stored in the beam before yield now scales as $\sigma_y^2/E^{4/3}$ with respect to material properties. For the material properties under consideration here, the plastic beam would be able to store 18.5 times the strain energy as the equal-stiffness steel beam before initiation of yielding. This in turn translates to a requirement of either 18.5 times the drop-weight mass or 18.5 times the height from which the mass is dropped to yield the plastic beam in comparison to the steel beam. For this particular example, a ductile engineering plastic would be much more dent resistant than a steel.

The energy balance approach to solving impact problems is not always as straightforward as this example. As a second, somewhat more involved example, consider the standard pendulum impact test for automotive bumpers. The applicable impact requirement in this standard is defined in terms of the bumper structure's survival of an impact by a test pendulum with mass equivalent to that of the car and a specific velocity [usually 4.0 or 8 km/h (2.5 or 5 mi/h)] prior to impact. The car in this test is specified to be in neutral. This requirement is independent of either the material or the geometry of the bumper structure. Such a requirement does not translate to a standard load history that can be applied to all bumpers interchangeably during design trade-off analyses. As illustrated in the general, two-body impact discussed previously, some bumpers will experience higher loads during the standard impact event because they are stiffer structural members. As a result, one cannot measure the force history associated with a steel bumper during a standard impact and simply apply it as a design load for a thermoplastic bumper.

There are a number of crucial questions to be addressed with respect to defining appropriate analyses that can be applied to efficiently design a bumper to withstand such an impact. Some of the critical questions that must be answered are the following:

1. How much energy must be absorbed by a bumper during such an impact?

2. What force level is developed on the bumper during such an impact?

3. How long does the actual impact event take and must the required analysis account for the inertial response of the bumper?

4. What magnitudes of strain rate should be expected in such a test and how much do they affect the load-carrying capability of the bumper?

Answers to these questions are important both from the standpoint of collecting appropriate material data as well as clearly defining mechanical analysis.

An initial model to be considered with regard to these basic questions is shown in Fig. 6.14 and discussed in Ref. 6. Here, the pendulum is modeled simply as a rigid solid of mass M with an initial velocity of V. The stiffness of the pendulum is assumed to be large in comparison to the plastic bumper. Likewise, the car is modeled as a rigid solid of equivalent mass M. Since the mass of the bumper is small with respect to the car, it is neglected in this analysis. Since the stiffness of a thermoplastic beam is likely to be small, however, it is included in the model as a spring of stiffness K_B. This stiffness can be quantified either experimentally or analytically by considering the displacement x of a bumper beam under a load F from a pendulum. Then the stiffness simply becomes

$$K_B = F/x \qquad (6.40)$$

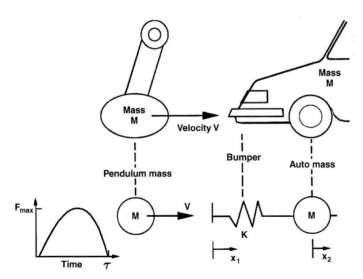

Figure 6.14 Lumped mass model for an automotive pendulum impact test for bumpers.

Although this is a very simplified model of the test, it provides a number of very basic pieces of information.

The governing differential equations for the dynamic system described in Fig. 6.14 are

$$M\ddot{x}_1 + K_B(x_1 - x_2) = 0 \tag{6.41a}$$

$$M\ddot{x}_2 + K_B(x_1 - x_2) = 0 \tag{6.41b}$$

and the initial conditions are

$$x_1(0) = 0 \tag{6.42a}$$

$$\dot{x}_1(0) = V \tag{6.42b}$$

$$x_2(0) = 0 \tag{6.42c}$$

$$\dot{x}_2(0) = 0 \tag{6.42d}$$

Solution of this set of simultaneous linear equations leads to

$$x_1 = \frac{V}{2}t + \frac{V}{2\sqrt{2}\,\omega} \sin \sqrt{2}\,\omega t \tag{6.43a}$$

$$x_2 = \frac{V}{2}t - \frac{V}{2\sqrt{2}\,\omega} \sin \sqrt{2}\,\omega t \tag{6.43b}$$

where

$$\omega = \sqrt{\frac{K_B}{M}} \tag{6.44}$$

and

$$t \le \frac{\pi}{\sqrt{2}\,\omega} \tag{6.45}$$

A number of very important conclusions can be drawn from Eqs. (6.43a) and (6.43b). First, an expression for the total force exerted by the pendulum on the bumper for $t \le \pi/\sqrt{2}\,\omega$ can be written as

$$F = K_B(x_1 - x_2) = V\sqrt{\frac{K_B M}{2}} \sin \sqrt{2}\,\omega t \tag{6.46}$$

As can be seen from Eq. (6.46), the time history of the impact force is a sinusoid with a maximum defined as

$$F_{\max} = V\sqrt{\frac{K_B M}{2}} \tag{6.47}$$

its deformation. As for other solids, the relationship between stress and strain rate takes the form[7]

$$\sigma_y = B_1 + B_2 \ln \dot{\varepsilon} \tag{6.58}$$

where B_1 and B_2 depend on the polymer and the temperature. In general, higher rates and lower temperatures lead to higher yield stresses, and lower rates and higher temperatures lead to lower yield stresses.

The temperature and strain-rate dependence of the yield stress for most polymers has been well investigated and reported in the literature. Data for a number of engineering thermoplastics such as polycarbonate[8,9] (PC), polyetherimide[9] (PEI), and polybutylene terephthalate[9] (PBT) are readily available in the literature. Figure 6.19 (Ref. 9) illustrates stress–strain relationships for polycarbonate, an amorphous polymer, at room temperature at a variety of strain rates (λ). The authors report that permanent deformation is observed when the maxima in the stress–strain curves are reached, and therefore they use this as their definition of yield stress. As can be seen in the figure, there is very little effect of strain rate evident in the lower stress regions of the curves. However, there is a distinct increase in yield stress for increasing strain rate. Figure 6.20 illustrates the strain-rate dependence of the yield stress for different temperatures. As can be seen, the yield stress also increases significantly as the temperature decreases.

Similar results[9] describing the stress–strain behavior of PEI, another amorphous polymer, are shown in Figs. 6.21 and 6.22. One interesting difference in the behavior of PEI is that there appears to be an

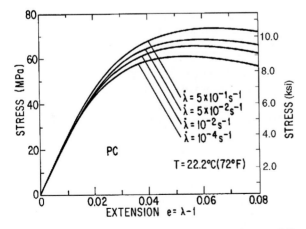

Figure 6.19 Stress–strain dependence of polycarbonate (PC) as a function of strain rate.

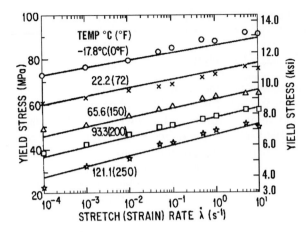

Figure 6.20 Strain-rate dependence of yield stress for polycarbonate (PC) at different temperatures.

observable difference in the initial stress–strain data, normally characterized by Young's modulus by engineers, for very slow strain rates. In contrast to PC and PEI, PBT is a semicrystalline polymer. Nonetheless, it displays behavior very similar to PC and PEI with regard to rate and temperature dependence of stress–strain behavior, as can be seen in Figs. 6.23 and 6.24.

The rate dependence of yield stress measured with standard tensile coupons reflects a true material property. As a result, these well-known

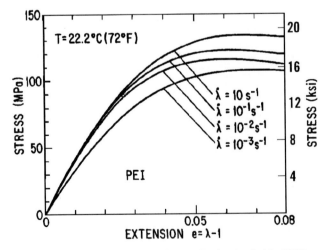

Figure 6.21 Stress–strain dependence of polyetherimide (PEI) as a function of strain rate.

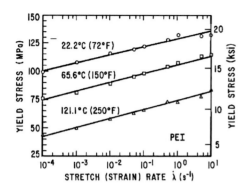

Figure 6.22   Strain-rate dependence of yield stress for polyetherimide (PEI) at different temperatures.

effects of rate can be incorporated in component analyses, either using simple closed-form equations or more precise numerical predictions often applied in design today. Once an approximate impact duration has been established, the strain rate in an impacted component can be approximated as shown in the example of the plastic bumper. With this approximation for strain rate and the temperature to which the component is exposed, data similar to that given in Ref. 9 can be used to establish an approximate yield stress for the material in this application. Clearly, the strain rate in a component will be different at different locations. However, since the effect of rate on yield stress is only significant over orders of magnitude in rate, an approximation of yield stress based upon maximum strain rate in a component is usually quite adequate for a first-order prediction of the component's elastoplastic response.

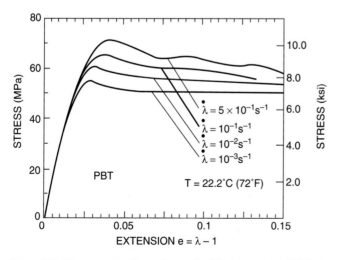

Figure 6.23   Stress–strain dependence of polybutylene terephthalate (PBT) at different temperatures.

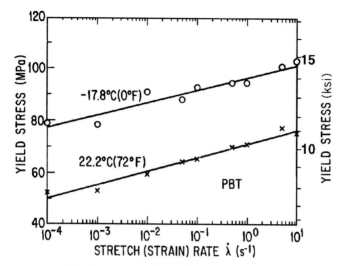

Figure 6.24 Strain-rate dependence of yield stress for polybutylene terephthalate (PBT) at different temperatures.

As an example of the accuracy that can be expected from making such an approximation, consider the boxlike structure shown in Fig. 6.25, discussed previously in Chap. 5 relative to plastic yield behavior. In the examples to be discussed here, this structure is subjected to dynamic loads at various rates through the bar at the top of the figure.[6] As will be recalled, this particular structure was molded from a rubber-modified polycarbonate, and the stress–strain response as a function of strain rate was measured in standard tensile tests and is shown in Fig. 6.26. Using a linear finite-element analysis, a relationship between the applied crosshead displacement rate and the maximum stress in the component was established. Next, elastoplastic analyses of the component were carried out using yield stresses associated with room temperature and the maximum strain rate in the component for the loading rate in question. Figures 6.27 and 6.28 illustrate the accuracy of this approximation in predicting the load–displacement behavior of the component and the maximum load sustained during the impact events as a function of strain rate. The maximum load sustained during these impact tests was associated with local regions of yielding and plasticity, as would be expected from the tensile test results. As can be seen, the correlation with experimental measurement is very good. It is also worth noting that for this material and component test, the rate effect on deformation was not extremely significant from a design standpoint. Standard low-speed, tensile test data would probably have been quite sufficient for design at the strain rates considered here.

these figures. This is an important design criteria for many bumper applications, since there are often constraints on the amount of deformation that a bumper can undergo before it begins to contact and damage other parts of the car. Clearly, adequate treatment of nonlinear issues, which are very often significant for plastic materials such as large displacements and yielding, allows very accurate prediction of the displacements associated with the impact of these plastic structures.

Rate effects and material behavior in the large-strain range

In the previous examples of the impacted box section and bumper beam, knowledge of the Young's modulus and yield stress as well as their dependence upon strain rate was sufficient for an accurate engineering analysis of the component. Even at strain rates representative of realistic impact events, these engineering properties are easily measured using standard tensile specimens. However, there are examples of much more severe deformation and damage to plastic materials due to impact. The puncture test, illustrated in Fig. 6.31, for example, subjects a plastic plate to strain and deformation much more severe than that experienced by the impacted box section. This test, which was previously discussed in Chap. 5, is one of the most regularly used measures of impact resistance for plastics. In comparison to strains of approximately 5% to 10% in the impacted box section, a punctured plate, such as the one shown in Fig. 6.32 and made of a ductile polymer, will have experienced true strains from 40% to 300%, depending on the material.

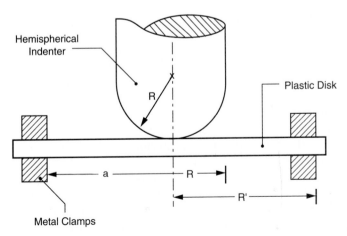

Figure 6.31 Schematic of a puncture test geometry.

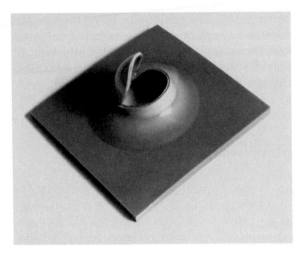

Figure 6.32 Punctured polycarbonate plate.

ASTM Procedure D3029 (Ref. 10) describes the geometry and test conditions recommended when using this specimen. The measured quantity most often quoted with respect to this test is energy required for failure. It should be emphasized that this quantity does not represent any fundamental material property. It cannot be used as part of a design analysis to predict the performance of a component with general geometry. It is only useful for comparing the performance of different polymers in this specific geometry. It is left to the qualitative judgment of the engineer to determine whether puncture resistance is a primary concern in the design of the component. Since puncture and drop dart tests are commonly used to rank plastic materials for impact resistance, a brief discussion of their mechanical behavior and relationship to material properties is merited.

First, let us consider whether it is necessary to consider any of the previously discussed inertial effects to understand the response of a plastic to a puncture test. Tabulations of natural periods of vibration for circular disks[11] indicate that for the standard geometry shown in Fig. 6.1, the lowest fundamental period is on the order of 2.35×10^{-4} s. The examples we examined earlier show that with loading times more than about 5 times the first fundamental period of a structure (1.2×10^{-3} s for this case), inertial effects of the structure will only result in variations of less than 10% from a static analysis. For a crosshead rate of 2.5 cm/s (1 in/s), the indenter load will be applied over a time on the order of a second. Even if a crosshead rate of 2.5×10^{2} cm/s (100 in/s) were applied, the load duration for this geometry would still be on the order of 0.01 s, which is still large enough to make inertial effects negligible. As a result, in spite of the fact that these tests are referred to as impact tests and rates of loading are high, the plastic disk's response is still quasistatic from an inertial point of view. A static mechanical analysis of the response will be highly accurate.

From a materials point of view, the performance of ductile polymers in a puncture or dart impact test is strongly affected by the material's stress–strain behavior at strain levels well beyond the yield range, as was discussed in Chap. 5, as well as Refs. 12 to 14. Of special significance is the fact that there is a range of behavior for many plastics at very large true strains (40% to 300% depending on the plastic) where substantial strain hardening occurs. Both the size of this hardening modulus as well as the true strain at which it becomes significant have been shown to be very important to the puncture resistance of polymers. Unfortunately, as previously discussed in Chap. 5, proper measurement of stresses and strains at such large deformation levels is very difficult. Consequently, there is much less fundamental material data in this range than there is with respect to elastic and yield behavior. The experimental work that has been done in this area[15–18] is predominantly at slow rates.

In their examination of the large-strain deformation of polycarbonate, polyetherimide, and polybutylene terephthalate, Nied and Stokes[18] provide some limited data relative to strain-rate effects upon large-strain deformation. Because their study was based upon tensile specimens that necked during the test, it was impossible to completely control the strain rate. However, they did briefly examine results of tests carried out at different crosshead rates. Figure 6.33 displays data

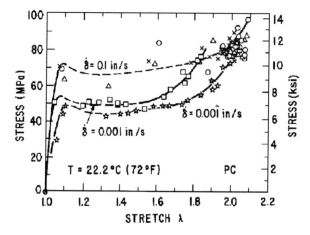

Figure 6.33 True stress versus stretch data for polycarbonate[18] at different crosshead rates.

relating true stress to stretch for polycarbonate in a tensile test at two different crosshead rates. Besides the previously discussed increase in yield stress, the authors note that they also observed an increase in "draw strain" for higher crosshead rates. This observation is most clearly displayed in Fig. 6.34, which reports the local stretch in a tensile coupon at one location as a function of crosshead displacement for different crosshead rates. The increase they note, however, is a modest one.

References 14 and 19 examine the first-order effects of approximating the large-strain behavior of polycarbonate as independent of temperature and strain rate. In these investigations, the simplified, trilinear approximation of stress–strain behavior in polymers, previously introduced in Chap. 5 and illustrated in Fig. 6.35, was used to examine the analytically predicted response of polycarbonate puncture disk coupons at different temperatures and strain rates. The yield-stress dependence upon temperature and strain rate was based upon the extensive data in Ref. 8. The large-strain material properties associated with this model, ε_d and E_3, were approximated using data in Ref. 18. Both ε_d and E_3 were assumed to be independent of temperature or rate.

The geometry of the axisymmetric puncture test is illustrated in Fig. 6.31. The most important dimensions governing the specimen's response are the indenter radius R, the specimen thickness t, and the unsupported disk radius a. Two different geometries were analyzed here. In the first geometry, R is 0.5 cm (0.2 in), a is 2.0 cm (0.8 in), and t is 2.5 mm (0.1 in). In the second, the indenter radius R is 1 cm (0.4 in),

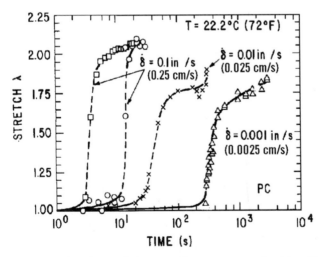

Figure 6.34 Local stretch in a tensile coupon as a function of crosshead displacement for different crosshead rates.

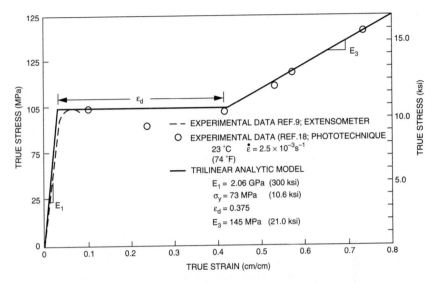

Figure 6.35 Trilinear large-strain constitutive model used in analyses.

the unsupported disk radius a is 2 cm (0.8 in), and the disk thickness t is 3.2 mm (0.125 in). For both geometries, the disks were clamped between two plates over the outside annuli extending from a to R' in Fig. 6.31.

Three comparisons between test results and finite-element analysis are illustrated in Fig. 6.36. In the first case, the 2.5 mm (0.1 in) thick disk was tested at room temperature and at a crosshead rate of 250 cm/s (100 in/s) using the 0.5 cm (0.2 in) radius indenter. In the second case, the test geometry and applied crosshead rate are identical to the first. However, the test was conducted at –60°C (–76°F). The third comparison offered here makes use of the 3.2 mm (0.125 in) thick disk and the 1 cm (0.4 in) radius disk. This test was carried out at room temperature and at a crosshead rate of 2.5 cm/s (1 in/s). All the analyses themselves are quasistatic in nature, since the loading rate is slow enough in comparison to the characteristic time periods for the disk to allow inertial effects to be neglected. Table 6.1 summarizes the pertinent geometry and material data defining these three comparisons.

Figure 6.36 compares the load–deflection curves predicted analytically with actual puncture test data. In all three cases, the correlation between analysis and experiment is reasonably good. As emphasized in Ref. 12, yielding in these puncture tests occurs very early in the test [below 2 mm (0.08 in) of dart deflection]. After the appearance of a geometrically induced stiffening effect due to large deflections and rotations, there is a noticeable softening in all three tests—an effect observed both experimentally and analytically. This reduction in the

TABLE 6.1　Table of Geometry and Test Conditions for Puncture Tests in Fig. 6.36

	R mm (in)	t mm (in)	a mm (in)	T °C (° F)	$\dot{\varepsilon}\,(\mathrm{s}^{-1})$	E_1 MPa (psi)	σy MPa (psi)	ε_0	E_3 MPa (psi)
1	5.0 (0.2)	2.5 (0.1)	20.0 (0.787)	25 (77)	10^2	2070.0 (300×10^3)	79.3 (11.5×10^3)	0.375	145.0 (21×10^3)
2	5.0 (0.2)	2.5 (0.1)	20.0 (0.787)	−60 (−76)	10^2	3000.0 (435×10^3)	120.6 (17.5×10^3)	0.375	145.0 (21×10^3)
3	10.0 (0.4)	3.2 (0.125)	20.0 (0.787)	25 (77)	10^0	2070.0 (300×10^3)	68.9 (10×10^3)	0.375	145.0 (21×10^3)

slope of the load–deflection curve occurs simultaneously with the appearance of a two-dimensional neck that initiates at the center of the disk beneath the indenter. However, in all three comparisons offered here, E_3 is large enough to arrest the initial neck and cause it to propagate outward. At the highest analytically predicted loads, the imposed indenter loads and displacements overcome the large strain material stiffening and a maximum is reached in the predicted load–deflection curve due to rapid thinning under the indenter head. The analytical predictions were terminated after maxima in the load–displacement curves that were associated with rapid thinning of the disk material in the general vicinity of the indenter. It is not clear that this analytically predicted thinning process is the same failure phenomenon that results in final rupture of the disk, but in the investigations reported here, the predicted maximum loads were always conservative.

Although these analyses are somewhat speculative in their nature, they illustrate the material properties that are significant to the performance of engineering thermoplastics in this test. As emphasized in Chap. 5, the stress–strain behavior in the large-strain range that is so significant to observed behavior in a puncture test has only recently been accurately measured, and the rate and temperature dependences of these relationships have not yet been defined. As a consequence, it should be emphasized that the drop-dart and disk-puncture impact tests are potentially quite complex in nature. In the most simple of situations, if the material is strictly linear elastic in its stress–strain behavior and failure is catastrophic once a limit stress is reached, the energy levels measured in this test will be quite low. For such materials the test is straightforward. However, for tougher plastics with larger strains to failure, this test includes significant nonlinearities, some purely geometry dependent and some directly related to material behavior.

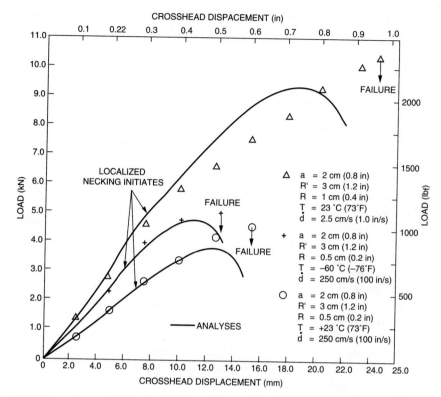

Figure 6.36 Comparison of predicted and measured load–displacement behavior of polycarbonate in puncture tests at different temperatures, strain rates, and test configurations.

Effect of strain rate and temperature on failure mode

The experimental behavior and related analyses discussed thus far with respect to impact events have all been associated with ductile material response and failure modes. Attention has been focused on the effect of rate upon the material properties that characterize this deformation process. Unfortunately, such behavior cannot be taken for granted in all engineering components. In some environments and configurations, even a polymer such as polycarbonate that is normally extremely ductile may fail in a brittle fashion. In Chap. 5, two significant factors contributing to this transition in mode of failure were discussed—temperature and stress state. If temperature is reduced sufficiently, the failure mode of a ductile polymer may change from ductile to brittle in nature. Furthermore, stress state can also play a significant role in transitions from ductile to brittle failure. Even at room temperature, stress states associated with large values of hydrostatic pressure (first invariant of the stress tensor), such as those induced in the vicinity of notches and other similar geometric details, will induce brittle failure in ductile polymers.

A third factor that can lead to brittle failure in ductile polymers is rate of loading. A plastic component that is normally ductile in nature at slow strain rates may fail in a completely brittle fashion at a relatively small strain if the rate of loading is increased sufficiently. Such behavior obviously has significant implications for a component where impact resistance is a design consideration. Whereas a ductile polymer at low strain rates will be tolerant of overloads because of its ability to redistribute load through the yielding process, at high rates, failure may be catastrophic in nature and associated with low energy-absorption levels.

Transitions from ductile to brittle failure as a function of temperature and strain rate are not peculiar to polymers. The technical history of this issue for metals is discussed by Timoshenko in Ref. 20. For metals, the transition from ductile to brittle failure is understood to be associated with two distinct deformation and failure modes: one producing brittle fracture by separation, the other corresponding to yield through sliding. In an entirely similar manner, Ward[21] uses two competing modes of deformation to discuss the ductile-to-brittle transition in failure for polymers.

The interrelationship of rate and temperature with regard to ductile-to-brittle transitions can then be illustrated qualitatively as shown in Fig. 6.37. Two sets of curves are shown in Fig. 6.37, one describing the brittle failure stress at two strain rates for a hypothetical polymer as a

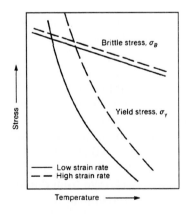

Figure 6.37 Qualitative illustration of strain rate and effects upon ductile-to-brittle transition temperature.

function of temperature, the other describing the yield stress of the same material for the same strain rates. The solid lines are associated with low strain rates and the dashed lines with high rates. The net effect illustrated in Fig. 6.37 is that as the strain rate is increased, the transition temperature defining the boundary between ductile and brittle failure moves to higher temperatures.

Practically speaking, transitions from ductile to brittle failure in tough engineering thermoplastics are most often associated with all three of the parameters discussed in the previous paragraphs—low temperature, high values of hydrostatic pressure (notchlike geometric configurations), and high strain rates. The notchlike geometric configuration is particularly important. It will be recalled from Chap. 5 that transition to brittle failure for polycarbonate in the puncture test geometry does not occur until almost −90°C (−130°F) even at strain rates on the order of 10 to 100 s^{-1}. However, a 6.4 mm (0.25 in) wide beam with a 0.127 mm (0.005 in) notch-tip radius will break in a brittle fashion at room temperature and strain rates as low as 10^{-3} s^{-1}. Unfortunately, since definition of the fundamental criterion governing brittle failure is not as well developed as the yield criteria defining onset of plastic flow, it is not currently possible to predict accurately the temperatures, geometric configurations, and loading rates associated with brittle failure in a general sense.

Because of the important effect of notch sensitivity on brittle failure at higher rates of loading, the Izod and Charpy tests are often used to characterize impact resistance of plastics. The Izod and Charpy tests, shown schematically in Fig. 6.38, are very similar in that they are both notched beam specimens subjected to bending moments. As discussed in Chap. 5, the notch serves to both create a stress concentration as well as produce a constrained multiaxial state of tension at a small distance below the bottom of the notch. Both of these effects tend to make the test severe from the standpoint of early transition to brittle behavior as a function of both rate and temperature. As a result, behavior in these tests is often quite different from disk puncture tests that are also used to assess impact resistance.

The Charpy geometry is a simply supported beam with a centrally applied load on the reverse side of the beam from the notch. The Izod geometry, on the other hand, is a cantilever beam with the notch located at the root of the beam. In both cases the load is applied dynamically by a free-falling pendulum of known initial potential energy. The important dimensions of interest for these tests include the notch angle, the notch depth, the notch-tip radius, the thickness of the beam, and the width of the beam. All of these quantities, as well as more detailed information specifying loading geometry and conditions, are specified in ASTM standards.[22] As the pendulum falls in both of these

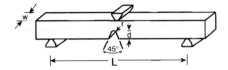

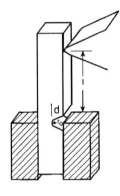

Figure 6.38 Charpy and Izod test configurations.

tests, its original potential energy is converted into kinetic energy. Some of this energy is in turn used to break the specimen, which is encountered at the low point of the pendulum arc. If the specimen is broken, the pendulum continues its swing and comes to a halt at a height less than its starting location. The energy expended to break the specimen can then be calculated as the difference between the initial and final potential energies of the pendulum. The value generally reported from these tests is the energy required to break the bar divided by the net cross-sectional area at the notch. The units of this measure are J/m^2 (ft · lb/ft^2) and the parameter is referred to as *impact strength.*

It should be emphasized that the units of this measure are not those of stress, nor is impact strength, as defined by these tests, a material property. In other words, the measurement cannot be used to design a component. In fact, the values of impact strength are significantly affected by the parameters defining the specimen geometry, such as notch-tip radius and beam width. Even the identification of a transition temperature can be significantly affected by geometry such as the width of the beam. Wider beams tend to provide more plane-strain constraint and transition temperatures often appear more distinct and at higher temperatures than results from thinner beams. As a result, when comparing materials via the value of impact strength it is imperative that the test geometries be identical. In addition, the transition temperature identified by plotting the impact strength as a function of temperature may or may not be appropriate for the component geometry under consideration.

Using the Charpy geometry, some of the qualitative effects of stress state, temperature, and rate can be illustrated with respect to polycarbonate. First, let us consider a 3.2 mm (0.125 in) thick, notched polycarbonate beam simply supported over a span of 102 mm (4 in) and loaded at midspan as shown in Fig. 6.39. The notch-tip radius in these tests was 0.25 mm (0.01 in). It will be recalled that this geometry was considered in Chap. 5 with respect to the relationship of crazing and brittle failure in notched geometries. For a loading rate of 5 mm/min (0.2 in/min) in that discussion, crazing was never observed and failure was always very ductile in nature with load–displacement curves similar to the ductile curve shown in Fig. 6.40.

If a servohydraulic testing machine is used to control the rate of loading, the maximum load achieved in this test configuration can be plotted as a function of crosshead speed, and the observed test results are presented in Fig. 6.41 over a range of crosshead rates from 2.5 × 10^{-4} to 1 × 10^2 cm/s (10^{-4} to 50 in/s). As can be seen in the figure, as the loading rate is increased, the failure load also increases. This behavior is consistent with the rate dependence of yield stress discussed earlier in this chapter. All of the failures are ductile.

Next, let us consider a notched beam with the same geometric configuration except that the beam thickness is increased to 6.4 mm (0.25 in). It will be recalled from the discussion in Chap. 5 that the thicker beam width led to higher values of hydrostatic pressure beneath the notch. Test results in this configuration at a crosshead rate of 5 mm/min (0.2 in/min) led to the observation of crazes forming beneath the notch tip. However, in spite of the appearance of crazes, the beam still exhibited a maximum in load–displacement behavior similar to the ductile curve in Fig. 6.40. If the maximum loads in this configuration are now plotted as a function of crosshead rate, the dependence illustrated in Fig. 6.41 is quite different from that illustrated for the 3.2 mm (0.125 in) thick beams in the same figure. Although the maxi-

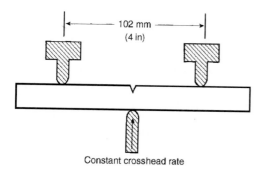

Constant crosshead rate

Figure 6.39 Notched-beam test geometry.

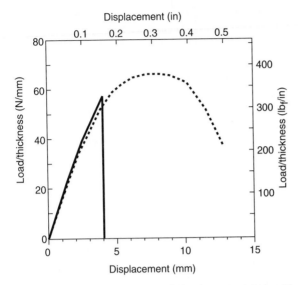

Figure 6.40 Load–displacement behavior associated with ductile and brittle failure of a notched polycarbonate beam.

mum load increases as the crosshead rate is increased from 2.5×10^{-4} cm/s (10^{-4} in/s), for crosshead rates beyond about 2.5 cm/s (1.0 in/s), the maximum in the load–displacement curve actually begins to decrease. In addition, there is more scatter in the value of maximum load and the lowest of these loads at a rate of 1.27×10^2 cm/s (50 in/s) is associated with failure characterized by the brittle load–displacement curve in Fig. 6.40. In contrast to the 3.2 mm (0.125 in) thick beams, the 6.4 mm (0.25 in) thick notched beams in this series of tests exhibit a transition from ductile to brittle failure mode as a function of rate.

In addition to geometry, temperature can also have an effect upon ductile-to-brittle transition as a function of rate. If the same 3.2 mm (0.125 in) thick beams discussed in relation to Fig. 6.41 are tested at −34°C (−30°F), then the dependence of maximum load upon crosshead rate is illustrated in Fig. 6.42. At low rates, the maximum loads for tests at −34°C (−30°F) are higher than the maximum loads for tests at a comparable rate at 23°C (73°F). This is due to the fact that yield stresses increase as temperature decreases. More importantly, although there was no ductile-to-brittle transition in failure as a function of rate observed in this configuration at room temperature, a transition as a function of rate clearly exists at −34°C (−30°F). Additional details of the rate and temperature effects upon failure for notched polycarbonate beams as well as comparable data for polyetherimide beams can be found in Ref. 23.

However, in the flexural test, linear-elastic, time-independent beam equations are used to calculate the bending stress using the applied load. Unfortunately, because of the time-dependent, nonlinear, stress–strain response of thermoplastics, the simple bending equations are often inadequate. Constant stress is not maintained because of stress redistribution—the stress distribution is not linear through the thickness of the beam. This phenomenon will be treated quantitatively later in this chapter.

A discussion of time-dependent measurement techniques must include a common measure of heat resistance: the heat deflection temperature (HDT). A bending specimen 127×12.7 mm (5×0.5 in), with a thickness ranging from 3.2 to 12.7 mm (0.125 to 0.5 in), is placed on supports 102 mm (4 in) apart, and a load producing an outer fiber stress of 0.46 or 1.82 MPa (66 or 264 psi) is applied. The temperature in the chamber is increased at a rate of 2°C/min (3.6°F/min). The temperature at which the bar deflects an additional 0.25 mm (0.010 in) is called the HDT or sometimes the distortion temperature under load (DTUL). Such a test, which involves variable temperature and arbitrary stress and deflection, is of no use in predicting the structural performance of a thermoplastic at any temperature, stress, or time. In addition, it can be misleading when comparing materials. A material with a higher HDT than another material could exhibit more creep than the other material at a lower temperature. Also, some semicrystalline materials exhibit very different values of HDT at 0.46 and 1.82 MPa (66 and 264 psi). For example, with polybutylene terephthalate (PBT), the HDT at 0.46 MPa (66 psi) is 154°C (310°F) and the HDT for 1.82 MPa (264 psi) is 54°C (130°F). The question of which HDT to use for comparison with another material that has the same HDT for both stress levels naturally arises. For purposes of predicting part performance and for material selection, tensile creep data is the desired measurement.

Another approach that is used to measure the change in material modulus with temperature is dynamic mechanical analysis (DMA).[2] A specimen is loaded with mechanical oscillation in bending or torsion at a fixed frequency. From the amplitude of the deformation and the applied load, the modulus is calculated. Such data are useful in showing how the modulus varies with temperature, allowing the identification of transitions, such as the glass transition.

Creep Models: Interpretation of Data

There is an extensive amount of information relative to creep models, data correlation and extrapolation, constitutive equations, and mathematical creep laws.[3–8] It is not possible to review all of the literature, but a few important techniques will be mentioned and then two examples of interpretations of creep data will be given. These examples focus on techniques leading to information that can be easily used in simple structural computational techniques or finite-element analysis programs.

The challenge to the design engineer is interpolation or extrapolation of the data to the particular time, temperature, and stress ranges of an actual application. This usually involves extrapolation in time and interpolation in temperature, both of which can be accomplished by using mathematical expressions or graphical, curve-fitting techniques. Time–temperature superposition is one technique of testing at a variety of temperatures and shifting the data on the logarithmic time scale with a shift factor. Essentially, testing is performed at higher temperatures for shorter times and shifted to produce information for longer times at lower operating temperatures. The use of a shift technique for data based on higher stress levels for shorter periods of time to lower stresses for longer periods of time is another approach. Examples of the use of mathematical expressions and graphical curve fitting will be presented.

Mathematical creep model

One of the primary concerns of a designer of polymeric components is excessive creep deformation. In order to be able to predict the deformation of components in the creep range, it is necessary to first study the factors affecting the accumulation of strain in a tensile specimen. This deformation under constant stress depends on three main parameters: time, stress, and temperature.[9] The most general creep equation is therefore

$$\varepsilon_c = f(t, \sigma, T) \tag{7.14}$$

where ε_c is the creep strain. A useful first approximation is to limit this general function to a commutative law of the form

$$\varepsilon_c = f_1(t)f_2(\sigma)f_3(T) \tag{7.15}$$

The separation of the functions $f_1(t)$ and $f_2(\sigma)$ has been implicit in most of the work on creep and appears to be generally acceptable for the purpose of calculations for components.[9] Separation of $f_3(T)$ is not as easily acceptable, since time and temperature have been combined into a single parameter by some investigators. However, an approximation of this kind with a separate function of temperature is reasonable in many cases.

The time dependence of creep under constant stress has received considerable attention and a number of expressions for it have been suggested. By choosing idealized mechanisms to represent the basic material behavior it is possible to derive various time functions. However, many of the functions have the form t^n where n is a constant. The function $f_2(\sigma)$ has been represented in many different ways. The most commonly used function is the power law. The reason for its popularity is its simplicity in application to stress analysis. Also, the power law is a close approximation to a hyperbolic sine function, which is also commonly used. For temperature changes where no change in material structure occurs, the most commonly used temperature dependence is of the form $\exp(-Q/RT)$, where Q is the activation energy, R is Boltzmann's constant, and T is the absolute temperature. Thus the expression that will be used for creep strain is

$$\varepsilon_c = a_0 t^n \sigma^m \exp\left(-\frac{Q}{RT}\right) \tag{7.16}$$

where a_0 is a constant.

The constants in the expression for creep strain in Eq. (7.16) were determined by graphical means for a 20% talc-filled polypropylene.[10] The total strain ε_t measured for constant uniaxial stress σ is assumed to be equal to the sum of the elastic strain, ε_e and ε_c, such that

$$\varepsilon_c = \varepsilon_t - \varepsilon_e \tag{7.17}$$

where $\varepsilon_e = \sigma/E$, and E is the elastic modulus. For a given temperature and applied stress, $\log \varepsilon_c$ is plotted versus $\log t$ so that n can be determined. For room temperature and $\sigma \leq 13.8$ MPa (2.0 ksi) and for 66°C (150°F) and $\sigma \leq 6.9$ MPa (1.0 ksi) n is about 0.3. For room temperature, n increases with stress above 13.8 MPa (2.0 ksi) with a value of about 0.4 at 17.2 MPa (2.5 ksi). Next, $\log \varepsilon_c$ is plotted versus $\log \sigma$ at constant time and temperature as shown in Fig. 7.4 where $t = 1000$ min and the slope m is about 2. Finally, when $\ln \varepsilon_c$ is plotted versus $1/T$ in Fig. 7.5 for $t = 1000$ min and $\sigma = 6.9$ MPa (1.0 ksi), Q is determined to be about 9 kcal/mol. Now, the expression for creep strain for 20% talc-filled polypropylene is

$$\varepsilon_c = 15.4\sigma^2 t^{0.3} \exp\left(-\frac{4560}{T}\right) \tag{7.18}$$

where σ is in MPa, t is in sec, and T is in K, or

$$\varepsilon_c = 0.0025\sigma^2 t^{0.3} \exp\left(-\frac{8200}{T}\right) \tag{7.19}$$

where σ is in psi, t is in min, and T is in °R. To test the overall fit of the creep equation to the data, $\log \varepsilon_c$ is plotted versus $\log t$ for 23°C (73°F) in Fig. 7.6, and for 66°C (150°F) in Fig. 7.7. Also, for 40% talc-filled

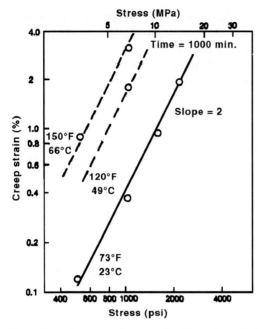

Figure 7.4   Creep strain as a function of applied stress for 20% talc-filled polypropylene.

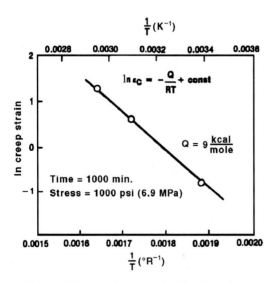

Figure 7.5   Creep strain as a function of inverse temperature for 20% talc-filled polypropylene.

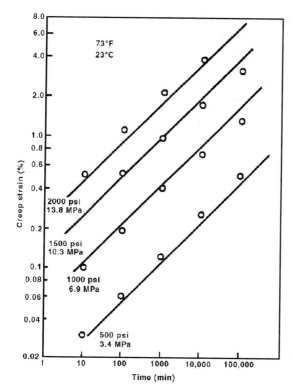

Figure 7.6 Comparison of creep law and tensile creep data at 23°C (73°F) for 20% talc-filled polypropylene.

polypropylene, the same creep equation can be used except that the coefficient is multiplied by 0.4, since polypropylene with 40% talc creeps only 0.4 as much as that containing 20% talc. In Fig. 7.8, log ε_c is plotted versus log σ for $t = 1000$ min to compare the creep equation for 40% talc-filled polypropylene with the creep data. In general, the creep equations provide a good fit of the data.

Graphical curve-fitting technique

Another interpretation of creep data, versus that of determining a mathematical creep model, is a more simple approach using standard curve-fitting techniques. Since load-bearing applications are designed for stiffness and strength, time–strain data are not directly useful. However, a curve-fitting interpretation of these data must retain the important features common to many of the creep models: log-time representation, power law for stress, and Arrhenius relations for temperature. In this section, the steps required to translate the initial

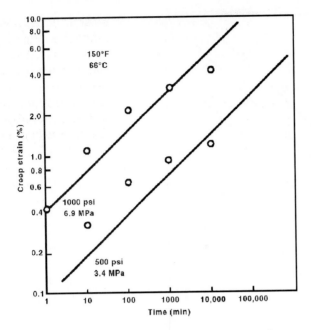

Figure 7.7 Comparison of creep law and tensile creep data at 66°C (150°F) for 20% talc-filled polypropylene.

strain–time data to an isochronous stress–strain curve for any time or temperature will be described.

Several methods for fitting and extrapolating time–strain data could be used. The objective should be to obtain the most accurate fit while achieving reasonable extrapolation predictions with minimum complexity. After considering a power law and an exponential law, it was determined,[11] based on visual inspection, that a slightly more complex representation would be necessary—a second-order polynomial function in log time where:

$$\varepsilon_t = A(\log t)^2 + B(\log t) + C \qquad (7.20)$$

The constants are determined by a least-squares curve-fitting method. An example[12] for polycarbonate resin tested at 20.7 MPa (3.0 ksi) and three temperatures is shown in Fig. 7.9. Engineering judgment must be used concerning the appropriate extrapolation in time. Caution should be exercised when more than one order of magnitude of time extrapolation is used. An example of the extrapolation prediction is shown in Fig. 7.10 where short-term (28-h) data for a polyetherimide (PEI) resin are extrapolated and compared with actual six-month creep

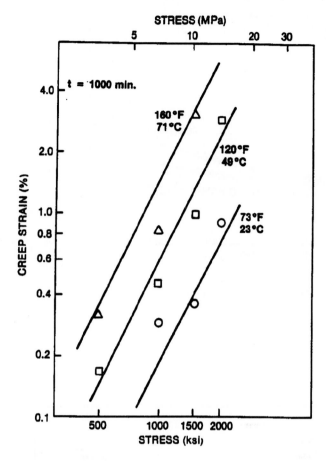

Figure 7.8 Comparison of creep law and tensile creep data for 40% talc-filled polypropylene.

data. The extrapolation is within 1% after one decade of time and 13.5% after two decades.

Based on fundamental principles for thermally activated processes, use of the Arrhenius relation is reasonable to interpolate and extrapolate with temperature. Plotting the strains versus temperatures for a particular time and stress on a natural logarithm (ln) of strain versus the inverse of the absolute temperature graph results in a linear interpolation or extrapolation based on the Arrhenius relation. For the polycarbonate resin tested at three temperatures (Fig. 7.9) and three stresses (with the creep data extrapolated to 1000 h), the Arrhenius plot is shown in Fig. 7.11 with the least-squares curve fitting. This shows a reasonable fit of the data. Again, caution must be exercised in

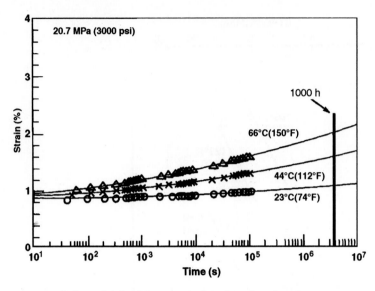

Figure 7.9 Polynomial fit of time–strain data for polycarbonate.

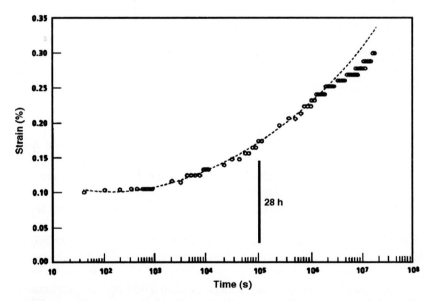

Figure 7.10 Comparison of extrapolated 28-h data (polynomial fit) to actual six-month data for a polyetherimide resin.

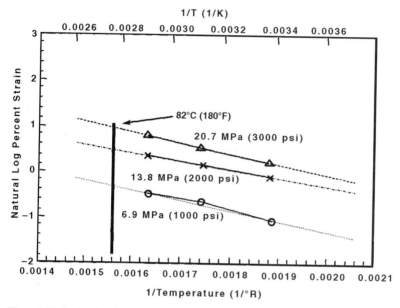

Figure 7.11 Arrhenius temperature extrapolation for time of 1000 h.

any temperature extrapolation to avoid a material structure change. For example, for a semicrystalline material such as PBT, this approach needs to be modified when the material is tested above its glass transition temperature (T_g). The temperature interpolation requires data from a minimum of two tests above T_g and two tests below T_g. Then the Arrhenius relation can be used above and below with different slopes (Fig. 7.12).

Finally, the isochronous stress–strain curve is produced by choosing the appropriate temperature and plotting the stress–strain points taken from the Arrhenius plots at that temperature and the previously chosen time. For example, for the polycarbonate resin for 1000 h and a temperature of 82°C (180°F) (Fig. 7.11), the isochronous stress–strain curve is produced (Fig. 7.13).

Time-Dependent Design Analysis

In this section, the previous creep models are used to predict time-dependent part displacements. First, the isochronous stress–strain curve is used to predict the time-dependent displacement of a CRT housing. Then, the mathematical creep models for polypropylene are used to predict creep in a plate with a hole and a beam in bending. Also. stress

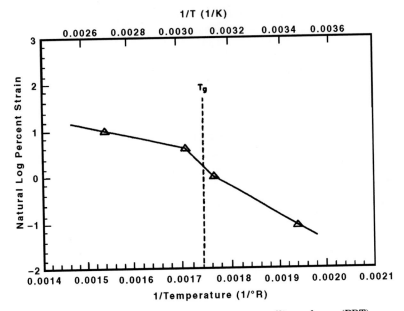

Figure 7.12 Temperature interpolation for a semicrystalline polymer (PBT).

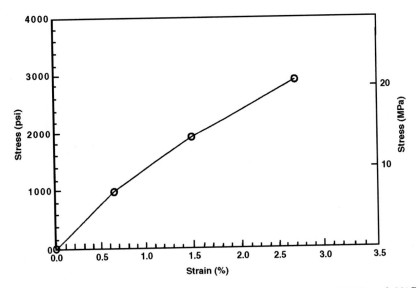

Figure 7.13 Isochronous stress–strain curve for polycarbonate for 1000 h and 82°C (180°F).

relaxation is considered for parts loaded to fixed displacement conditions, such as press fits.

For constant stress applications, the isochronous stress–strain curve can be used with standard equations by choosing the appropriate "effective modulus" considering the range of stresses in the application.[12] This requires engineering judgment where higher stressed parts would typically be analyzed with a lower "effective modulus." For example, Fig. 7.14 shows three different moduli that might be chosen depending on whether the estimated maximum stress was about 6.9, 13.8, or 20.7 MPa (1.0, 2.0, or 3.0 ksi). When the isochronous stress–strain curve is highly nonlinear or the part geometry is complex, finite-element structural analysis techniques can be used. Then, the complete nonlinear, isochronous stress–strain curve can be used in a nonlinear finite-element analysis or a linear effective modulus can be used in a linear analysis.

In order to assess the general applicability of this method for predicting the creep deformation of a part, analytical predictions for a simple geometry were compared to experimental results.[12] The component used was a polycarbonate CRT housing (Fig. 7.15). The housing was clamped on all three sides at the base and loaded vertically by a 12.7 mm (0.5 in) radius dart at the top center. The finite-element model is shown in Fig. 7.15 with appropriate boundary conditions. The load could be effectively modeled by a distributed load on four nodal points. Tests were performed at 82°C (180°F) to measure the deflection with

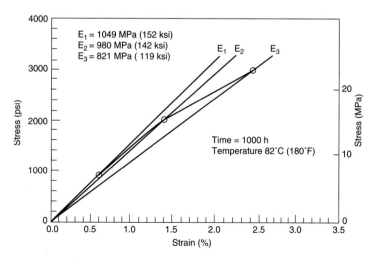

Figure 7.14 Three effective moduli obtained from one isochronous stress–strain curve.

(a)

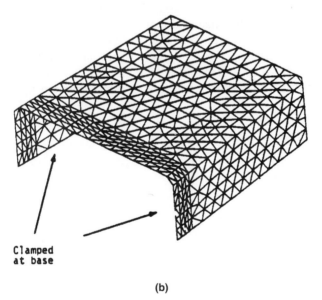

Clamped
at base

(b)

Figure 7.15 (a) Photograph of actual CRT housing tested. (b) Finite-element representation of the CRT housing used in analysis.

time. Using the data for polycarbonate described earlier, isochronous curves could be produced for various times to 100 hours (Fig. 7.16). Using these stress–strain curves, finite-element analyses were performed and the results were compared with the experimental results in Fig.

Variable Loading and Recovery

The time-dependent behavior of plastic parts considered so far in this chapter has been limited to constant loading. However, components are usually subjected to a complex pattern of loading and unloading situations. It is impossible to generate material data for each loading situation.[13] Design analysis based on assumptions of constant loading at the maximum load would be overly conservative. Simple methods of accounting for variable loading and recovery must be used. The simplest workable theoretical model proposed to predict the strain response to a stress history is the Boltzmann superposition principle.[1] For this principle the strain response for any complex loading history is simply the algebraic sum of the strains accumulated for each load step. Thus, the behavior of the plastic is a function of the entire load history.

When a polymeric material is subjected to a constant stress, σ_0, at time zero, the time-dependent strain is expressed as

$$\varepsilon_0(t) = \frac{\sigma_0}{E(t)} \qquad (7.24)$$

where $E(t)$ is the time-dependent modulus. $E(t)$ can be determined by examining the isochronous stress–strain curve such as Fig. 7.16 or by dividing the stress by the associated time-dependent strain $\varepsilon(t)$ from Eq. (7.20). For an additional stress σ_1 applied at a time t_1, the strain is given by

$$\varepsilon_1(t) = \frac{\sigma_1}{E(t - t_1)} \qquad (7.25)$$

If σ_0 was applied at time zero and an additional stress σ_1 was applied at time t_1, then Boltzman's superposition principle states that the total strain is the sum of the two responses:

$$\varepsilon(t) = \varepsilon_0(t) + \varepsilon_1(t) \qquad (7.26)$$

Furthermore, if the total stress $(\sigma_0 + \sigma_1)$ is removed at time t_2, the strain is

$$\varepsilon(t) = \varepsilon_0(t) + \varepsilon_1(t) + \varepsilon_2(t) \qquad (7.27)$$

where $\varepsilon_2(t) = -(\sigma_0 + \sigma_1)/E(t - t_2)$. The example given in Fig. 7.20 is for polycarbonate at a temperature of 66°C (150°F). Note that σ_0 and $\sigma_1 = 6.9$ MPa (1.0 ksi), $t_1 = 1$ h, and $t_2 = 2$ h.

This principle can be generalized as

$$\varepsilon(t) = \sum_{i=0}^{N} \frac{\sigma_i}{E(t - t_i)} \qquad (7.28)$$

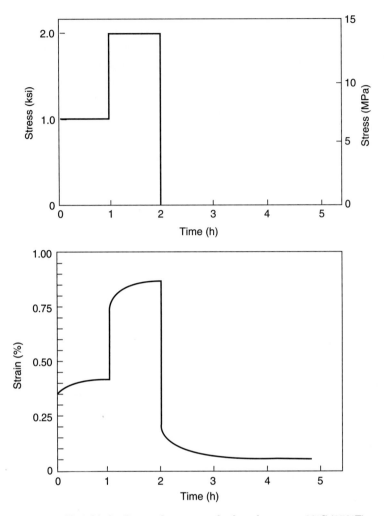

Figure 7.20 Variable loading and recovery of polycarbonate at 66°C (150°F).

where N is the number of step-load changes and σ_i is the additional stress applied at time t_i. Similarly, for stress relaxation under constant strain conditions, ε_i,

$$\sigma(t) = \sum_{i=0}^{N} \varepsilon_i \, E(t - t_i) \qquad (7.29)$$

A simple example of loading to a constant stress σ_0, for a time t_1, and then unloading illustrates a simple recovery process. If the relative recovery time is treated as a total time divided by the creep time, t/t_1, and the relative recovery strain $\varepsilon_r(t)$ as the strain for $t > t_1$ divided by the time-dependent strain $\varepsilon(t_1)$ then

$$\varepsilon_r(t) = \frac{\dfrac{\sigma_0}{E(t)} - \dfrac{\sigma_0}{E(t-t_1)}}{\dfrac{\sigma_0}{E(t_1)} - \dfrac{\sigma_0}{E(0)}} \tag{7.30}$$

The relative recovery strain [Eq. (7.30)] is plotted versus the relative recovery time in Fig. 7.21 for polycarbonate. Also, the mathematical expression developed for the creep strain for polypropylene for a given stress σ_0 and temperature is equal to a constant times $t^{0.3}$. Thus, substitution into the relative recovery expression yields

$$\varepsilon_r(t) = \frac{t^{0.3} - (t-t_1)^{0.3}}{t_1^{0.3}} = \left(\frac{t}{t_1}\right)^{0.3} - \left(\frac{t-t_1}{t_1}\right)^{0.3} \tag{7.31}$$

This result is plotted in Fig. 7.21.

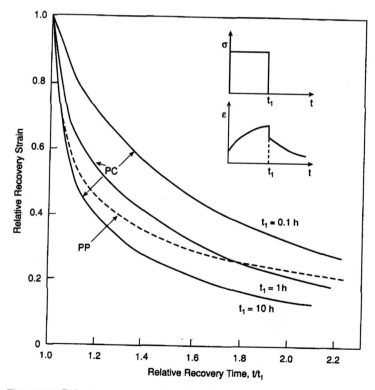

Figure 7.21 Relative recovery strain (strain for $t > t_1$ divided by creep strain at $t = t_1$) for polycarbonate (PC) and polypropylene (PP).

Actually, the primary concern to a design engineer is the time-dependent deflection of a structure subjected to a variable load history. The previous concepts and equations can simply be extended by performing an elastic analysis to establish the geometry constant k where E is the elastic modulus, δ is the part displacement under a load P, and

$$\delta = \frac{kP}{E} \tag{7.32}$$

By analogy,

$$\delta(t) = k \sum_{i=0}^{N} \frac{P_i}{E(t - t_i)} \tag{7.33}$$

Thus, Eq. (7.33) can be used with an elastic part analysis to predict part displacement with variable loading (P_i).

Time-Dependent Rupture

Certainly one of the prime concerns of the design engineer is excessive deformation of the component. However, in some cases, rupture must also be considered. Materials that exhibit time-dependent deformation must eventually rupture. If significant deformation precedes rupture, the function of the component will probably be impaired prior to rupture. However, for some materials subjected to high stresses, rupture must be considered. The objective here is to provide a general treatment of material rupture behavior. The time for the material to fracture is dependent on the applied stress, the ambient temperature, other environmental conditions, the component geometry, and the fabrication process. The design engineer must have an awareness of creep rupture and use available data as guidelines since a proven methodology does not exist.

An example of the relative importance of deformation and rupture[14] is illustrated in Fig. 7.22. The isometric curves (constant strain) can be taken from creep data at constant temperature and various constant stress levels or from isochronous stress–strain curves. In this case, relatively large strains would occur prior to ductile creep rupture. Of concern is brittle rupture occurring after relatively small levels of strain. Also, some materials that may exhibit ductile rupture at short periods of time may exhibit brittle failure after longer periods of time[15]. Polyethylene used in pipe applications exhibits this behavior, for example. Examination of the isometric curves and the creep rupture curve is the best indicator of brittle creep rupture. If the curves are parallel, ductile failure primarily dominated by strain could be indicated. Ob-

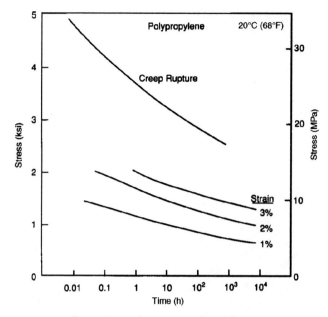

Figure 7.22 Comparison of creep rupture with isometric stress for polypropylene.

servation of the fracture line intersecting the isometric lines would indicate brittle rupture.

The effect of temperature on time-dependent rupture is an important concern as well. The 1000-h stress rupture data for filled and unfilled thermoplastics shown in Fig. 7.23 provide an overview of the effect of temperature.[16] The purpose of the linear–linear plot is to demonstrate the general trends and not to imply any particular mechanism.

The mechanism of time-dependent rupture is certainly not well understood. A fracture mechanics interpretation would involve initiation and propagation of a crack as a damage accumulation process leading to rupture. A number of investigators have developed various methods to account for this phenomenon.[17,18] The basis of these methods is that creep damage under constant conditions is proportional to the fraction of total rupture life under those conditions. Also, the damage resulting from each loading period is independent of all other loading periods at different stresses and temperatures. The total damage is the summation of damage from each loading period:

$$\sum_{i=1}^{N} \frac{\Delta t_i}{t_{ri}} = 1 \qquad (7.34)$$

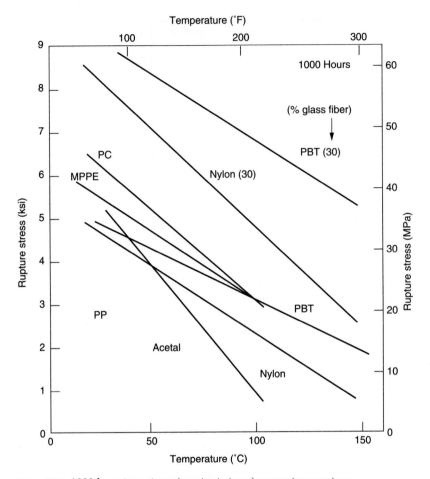

Figure 7.23 1000-h rupture stress (constant stress) versus temperature.

where N is the number of loading periods (load conditions), Δt_i is the time of the period, and t_{ri} is the rupture time for the loading period. In integral form, rupture for variable temperature and stress takes the form

$$\int \frac{dt}{t_r} = 1 \tag{7.35}$$

The time to rupture for constant stress loading of polypropylene specimens is shown in Fig. 7.24 as a log–log plot of applied stress versus time to failure. The reasonable fit of the data indicates that a power-law relationship of the type

$$t_r = k\sigma^{-n} \tag{7.36}$$

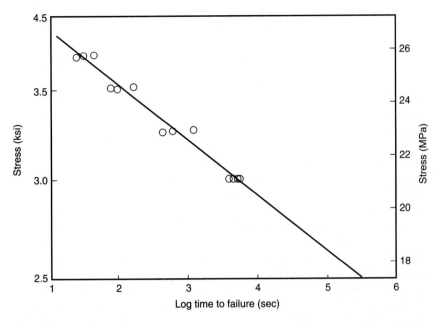

Figure 7.24 Creep rupture (time to failure for constant stress) of 20% talc-filled polypropylene at room temperature.

is appropriate, where k and n are constants ($n = 24$). Substitution of Eq. (7.36) in Eq. (7.35) gives

$$\int_{0}^{t_f} \sigma^n \, dt = k \tag{7.37}$$

A similar relationship can be derived by assuming that time-dependent crack growth controls the fracture behavior of polymers. For this process, the crack velocity during crack propagation depends primarily on the stress intensity factor (K), where A and n are constants that depend on temperature and environment, and

$$\frac{da}{dt} = AK^n \tag{7.38}$$

The crack growth rate expected under any loading can be evaluated for the condition that the growth rate is controlled by Eq. (7.38). Substituting the expression for the stress intensity factor,

$$K = Y\sigma \sqrt{a} \tag{7.39}$$

where Y is a crack geometry factor and a is crack length in Eq. (7.38) results in

$$\int_{a_i}^{a_f} \frac{da}{a^{n/2}} = AY^n \int_0^{t_f} \sigma^n \, dt \tag{7.40}$$

where the time to failure t_f consists of the time for the crack to grow from an initial size a_i to a final, critical size a_f. Since a_i, a_f, A, Y, and n are constants, Eq. (7.40) reduces to Eq. (7.37). Thus, a fracture mechanism of time-dependent crack growth and the general creep damage mechanism yield the same simple relationship.

Data for polystyrene[19] can be used to illustrate (1) the effect of an environment—methanol, (2) the correspondence between constant stress and constant stress-rate loading, and (3) the difference between craze initiation and rupture. Constant stress and constant stress-rate loading data should have the same slope (n), where log t_r is plotted versus log σ with only the coefficient changing [Eq. (7.36)]. Figure 7.25 shows a plot of craze initiation stress and fracture stress as a function of stressing rate where the difference between craze initiation and creep rupture is clearly displayed. However, for the methanol environment, the slope n for craze initiation and rupture under constant stress-rate loading and for constant stress loading is nearly the same (Table 7.1). In ambient air, the n value is significantly higher but nearly the same for rupture and craze initiation under constant stress-rate loading (see Table 7.1). These examples demonstrate the type of

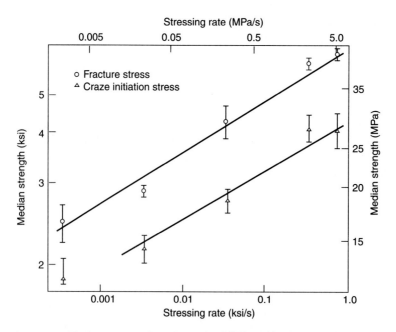

Figure 7.25 Fracture strength and craze initiation stress of amorphous polystyrene in methanol as a function of stressing rate.

TABLE 7.1   Slope (*n*) of the Time-Dependent Fracture Representation of Polystyrene

	Constant stress rate		Constant stress
	Rupture	Craze initiation	
Methanol	7.52	7.66	7.43
Air	24.25	25.35	...

available data that design engineers can use to compare with component stresses as a guideline for considering the potential of creep rupture in a component.

References

1. J. J. Aklonis and W. J. MacKnight, *Introduction to Polymer Viscoelasticity,* Wiley, New York, 1983.
2. M. P. Sepe, "Material Selection for Elevated Temperature Applications: An Alternative to DTUL," *Proceedings of the 1991 Society of Plastics Engineers (SPE) Annual Technical Meeting,* SPE, Brookfield Center, CT, 1991, pp. 2257–2262.
3. T. Sterrett and E. Miller, "A Comparison of Empirically and Theoretically Based Functions for the Modeling of Plastics Creep Data," *Proceedings of the 1984 Society of Plastics Engineers (SPE) Annual Technical Meeting,* SPE, Brookfield Center, CT, 1984, pp. 495–497.
4. E. Reichelt, "Long-Term Deformation Behavior of Plastics," *Kunststoffe 76* (10): 971–974, 1986.
5. W. N. Findley, *An Introduction to Linear Viscoelasticity,* North-Holland, Amsterdam, 1976.
6. H. F. Rondeau, "You Can Predict Creep in Plastic Parts," *Modern Plastics,* March 11, 1976.
7. J. R. McLoughlin, "A New Creep Law for Plastics," *Modern Plastics,* February, 97, 1968.
8. G. S. Brockway, "Taking the Mystery Out of Creep," *Plastics Design Forum,* January, 25, 1982.
9. R. K. Penny and D. L. Marriott, *Design for Creep,* McGraw-Hill, London, 1971.
10. G. G. Trantina, "Creep Analysis of Polymer Structures," *Journal of Polymer Engineering and Science,* 26 (July): 776, 1986.
11. G. G. Trantina and D. A. Ysseldyke, "An Engineering Design Database for Plastics," *Materials Engineering,* 104 (October): 35–38, 1987.
12. D. A. Ysseldyke and J. Messaros, "Utilizing Computer-Aided Engineering in Predicting Structural Creep Deformation, *Proceedings of the 1987 Society of Plastics Engineers (SPE) Annual Technical Meeting,* SPE, Brookfield Center, CT, 1987, p. 1480.
13. M. J. Mindel, "Creep and Recovery of Polycarbonate," *Journal of Material Science,* 8: 863–870, 1973.
14. R. M. Ogorkiewicz, *Engineering Properties of Thermoplastics,* Wiley-Interscience, London, 1970.
15. R. J. Crawford, *Plastics Engineering,* Pergamon, Oxford, 1987.
16. *Modern Plastics Encyclopedia,* McGraw-Hill, New York, 1988.
17. E. L. Robinson, *Transactions of the ASME,* 60, 1938.
18. R. M. Goldhoff, "Uniaxial Creep Rupture Behavior of Low Alloy Steel Under Variable Loading Conditions," *Transactions of the ASME, Journal of Basic Engineering,* 87, 1965.
19. J. E. Ritter, J. M. Stevens, and K. Jakus, "Failure of Amorphous Polystyrene," *Journal of Material Science,* 14: 2446–2452, 1979.

Fatigue: Cycle-Dependent Part Performance

An understanding of the deformation and fracture behavior of plastics subjected to cyclic loading is necessary in order to predict the probable lifetime of structures fabricated from plastics. This fatigue behavior is of concern since failure at fluctuating load levels can occur at much lower levels than failure under monotonic loading. A significant amount of information exists on the fatigue behavior of plastics.[1-4] Unfortunately, there has been very little documented about the application of this understanding to predict the fatigue behavior of plastic parts.

There are two distinct approaches to treating and measuring the fatigue of polymers. The first approach is the traditional measurement of the number of cycles to failure (N) as a function of the fluctuating load or stress (S)—(S-N). The "load" that is controlled is the minimum and maximum force or displacement in tension or bending. The fluctuations have a certain frequency and waveform. From a design viewpoint, it is difficult to predict part performance with these data because an enormous number of variables must be taken into consideration as well as various environmental conditions and a wide variety of materials.

The second approach to treating the fatigue of plastics is cyclic crack propagation. The use of fracture mechanics in cyclic fatigue involves the measurement of the amount of crack growth per cycle as a function of the stress-intensity factor. The fundamental addition here is the treatment of the crack length and thus an improved understanding of a fatigue mechanism. However, the same large number of variables that apply to the traditional fatigue (S-N) approach apply to the crack propagation approach. In addition, the design engineer is challenged with determining the initial or inherent flaw size.

Even though cycle-dependent part performance is not well understood, a general design engineering approach can be applied to the fatigue of plastic parts. First, for material selection, an awareness of the fatigue performance of numerous plastics is necessary. Materials should be compared under identical test conditions to determine their relative fatigue performance. Knowledge should be gained on the effects of toughening and the addition of particulate and fiber reinforcements. This preliminary selection should be based on the general assessment of the relative fatigue performance, taking into account the overall severity of the part loading. Next, the part loading conditions should be determined and related to the appropriate laboratory data. This task is probably the most important, yet the most difficult, due to the range of variables involved. Establishing whether the part will experience load-controlled or displacement-controlled cyclic loadings is possibly the most significant factor. Next, the effects of frequency, waveform, and load level and type must be assessed to determine if the part temperature will increase, leading to thermal fatigue, or if mechanical failure will occur with little or no temperature increase. Other conditions that should be considered or matched from the laboratory specimen to the component include environmental effects, e.g., temperature, stress state, stress concentrations, and mean stress. Finally, depending on the engineer's ability to achieve the second task, appropriate laboratory tests or full-scale component tests should be conducted. These laboratory tests must be carefully planned in order to achieve correspondence to the actual service conditions.

In this chapter the emphasis will be on understanding the influence of important loading variables so component lifetimes can be assessed more accurately. Following a general presentation of S-N data for various thermoplastics, loading variables such as control mode (load versus deflection), frequency, waveform, mean stress and stress concentration, and stress state will be considered. Then, cyclic crack propagation will be presented. Finally, some examples of fatigue lifetime prediction for components will be included.

General Load–Lifetime (S-N) Fatigue Representation

The fatigue lifetime (number of cycles to failure) of a part is strongly dependent on the applied load. Representative S-N curves for a number of thermoplastics[5] are shown in Fig. 8.1. All of these data were generated at room temperature with a standard tensile specimen with a net cross section of 12.7×3.2 mm (0.5×0.125 in). The tensile load was

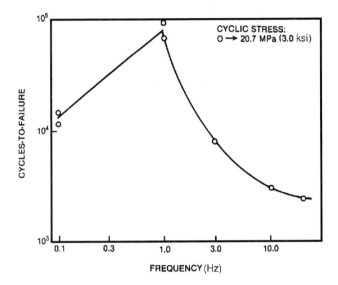

Figure 8.4 Lifetime as a function of cyclic frequency—self-heating above 1 Hz.

produce a larger temperature rise because the heat must be conducted through the material to its surface before it can be transferred to the surroundings by convection.

An example of the effect of frequency on fatigue lifetime[6] is shown in Fig. 8.4 for polypropylene subjected to stress-control cycling from 0 to 20.7 MPa (3.0 ksi). Above 1 Hz the lifetime decreases because of hysteretic heating leading to thermal failure. For 30 Hz cyclic loading, the surface temperature of the specimen increases 30°C (86°F), while for 1 Hz the temperature increase was only a few degrees. However, for frequencies below 1 Hz the lifetime also decreases, but for a different reason: creep–fatigue interaction, to be described later in this chapter.

Another example of the transition from mechanical to thermal fatigue is shown in Fig. 8.5 for acetal.[9] Figure 8.5a shows the thermal runaway for the high stress levels and the temperature stabilization at decreasingly lower temperatures with lower stress levels. Figure 8.5b indicates the transition from thermal failures to mechanical failures with decreasing stresses, and longer lifetimes with decreasing frequencies (similar to polypropylene for frequencies ≥1 Hz).

Hysteretic heating is an important consideration in the fatigue failure of plastic components. The designer should attempt to minimize heat generation and to promote heat transfer if hysteretic heating is identified as a failure mechanism. Stresses can be reduced by adding ribbing, flanges, or other geometry changes to distribute the stresses. If possible, the frequency should be reduced or plastics with a low

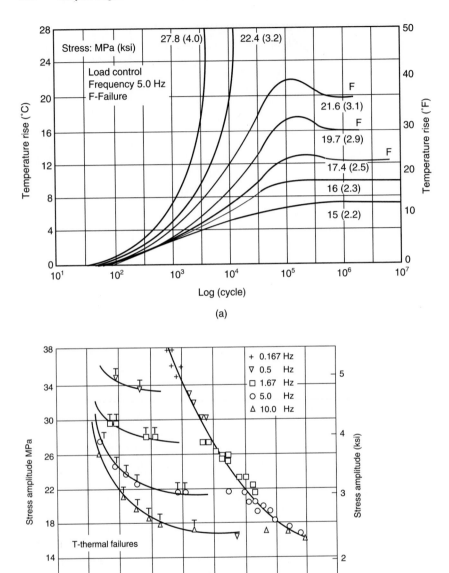

Figure 8.5 (*a*) Temperature rise during cyclic loading in acetal. (*b*) Typical fatigue behavior of acetal at several frequencies.

loss compliance should be selected. Polysulfone (PSO) and modified polyphenylene ether (M-PPE) have a low loss compliance; polycarbonate, acetal, and polyethylene terephthalate (PET) have a medium loss compliance; and polypropylene, polyethylene, and nylon have a high loss compliance. Heat transfer can be increased by decreasing wall thicknesses, adding fins, and providing air circulation.

Cyclic Waveform

Cyclic waveform can have an influence on fatigue lifetime depending on the fatigue failure mechanism. The most common waveform is sinusoidal, but the square wave and the ramp (linear) waveform represent extremes that can bound the effect on fatigue lifetime. For thermal softening failures, the square wave results in maximum energy dissipation per cycle and thus the lowest stress amplitude where thermal fatigue occurs. For the ramp waveform, there is a minimum of energy dissipation per cycle and thus higher stresses are possible before thermal fatigue occurs. The sinusoidal waveform represents an intermediate level of energy dissipation. The cyclic waveform can also have an influence on the creep–fatigue failure mechanism and the crack growth-rate failure process. These situations will be discussed relative to the effect of waveform later in this chapter.

Relationships between Creep Rupture and Fatigue

For some polymers there is a relationship between creep rupture and fatigue failure. One approach to the prediction of cyclic fatigue based on the use of creep rupture data[6] will be presented here and demonstrated for mechanical fatigue failure of polypropylene (frequency ≤ 1 Hz). When this fatigue lifetime prediction scheme is applicable, the design process is simplified since only creep rupture data are needed to predict the effects of cyclic frequency and waveform.

In the preceding chapter on time-dependent behavior, creep rupture data were treated with a power-law relationship of applied stress and time to failure. A treatment of the failure process for any stress history was developed as a damage summation process or a crack growth-rate process [Eq. (7.37) in the preceding chapter]. The next step in predicting the number of cycles to failure involves substituting the cyclic stress history into this expression. For a triangular waveform, the stress is linearly proportional to time, while for a sinusoidal waveform, the stress–time relation is a sine function. In general, the result is

$$\frac{\sigma^n t_f}{f(n)} = k \qquad (8.2)$$

where σ is the maximum applied stress, the minimum applied stress is zero, and $f(n) = n + 1$ for a triangular waveform and $f(n) = 8.7$ for a sinusoidal waveform with $n = 24$. For the triangular waveform, the integration is straightforward, while for the sinusoidal waveform, a series solution is possible when n is an integer.[10] In any case, $f(n)$ represents the ratio of the times to failure for cyclic and constant stress loading to the same stress level. Since

$$t_f = \frac{N_f}{\nu} \qquad (8.3)$$

where N_f is the cycles to failure and ν is the cyclic frequency,

$$N_f = k\nu\sigma^{-n}f(n) \qquad (8.4)$$

Thus, the cyclic lifetime as a function of frequency and waveform can be predicted from the creep rupture data.

For 20% talc-filled polypropylene, the fatigue data shown in Fig. 8.6 for 1 Hz sine and triangular waveforms are predicted very accurately by Eq. (8.4). For 0.1 Hz (Fig. 8.7), the prediction is conservative by about a factor of 2 in fatigue lifetime. Overall, cumulative damage models have been used with mixed success depending on the polymer. Two of three types of polyethylene were adequately described by a cumulative damage model.[11] A methodology for describing creep–fatigue interactions in thermoplastic components was applied to polyethylene pipes, demonstrating the need for nonlinear, additive damage.[12] Polyethylene and PMMA data were shown to deviate from predictions in different ways.[13] Also, fatigue and creep rupture data for ABS (Ref. 14) show that, at equivalent time under tensile load, the square-wave loading is more damaging than a static load (Fig. 8.8). These data also demonstrate a distinct slope change in the fatigue data, with ductile failures occurring at the higher stresses and hysteretic heating and brittle, surface-initiated failures occurring at lower stresses. Data[14] for another ABS with a different rubber content show a shift on the stress axis, but the two curves coincide by normalizing the applied stress with respect to the yield stress.

Stress Concentrations

Stress concentrations are an important cause of fatigue failure in components. In addition to the effect of the change in geometry of components causing stress concentrations, there are also many concentrations resulting from the fabrication process, such as sprue marks, weld lines, and surface finish. Stress concentrations combined with the ten-

9

Closure

Despite their large volume usage, polymers are still a new class of engineering materials. And as previously discussed, several categories of supporting technology must evolve as the usage of any new material matures. First, a fundamental understanding of the elemental structure of the material must take place in order to facilitate material improvement and invention and, in turn, broaden the variety of material choices. In addition, processes used for forming the material into useful shapes must also be identified and refined to ensure its practical application. Such technologies foster the emergence of the initial commercial concepts for use of the material.

At this point in time, the fundamental groundwork in chemistry, materials science, and process engineering for polymers has been laid. In many, but not all, situations, mechanical properties are sufficiently well known and understood to consider the use of these materials in load bearing applications. Now, engineering design and analysis technologies relevant to these materials must be further developed in order to support the continued growth in use of these materials for more challenging applications. As the role of plastics in load-bearing applications expands, it will become ever more important to perform accurate engineering analyses and optimize plastic part performance. The current status of the development of these structural design analysis techniques has been the subject of this book. While there are still gaps in the knowledge base for these materials, application of the technology that does exist represents a significant improvement over the time-consuming and costly build-and-test approach that has previously been used.

The primary focus of this book is to document the application of modern analysis techniques to facilitate the design of structural parts made of thermoplastic materials. Although these methods are an important element in design, the text has not attempted to provide a general procedure for the design of plastic parts. However, what the book does concentrate on is the application of material properties, within the context of mechanical analysis techniques, to predict plastic part performance. In spite of the significant amount of knowledge and methodology associated with plastics that has evolved over the last 10 to 20 years, there are still many areas where additional development would be a great benefit. As a partial response to that ongoing need, the objective of this chapter is to reflect on the general content of this book and suggest areas where technology still needs to be refined.

In Chap. 4 of this text, the general subject of plastic part stiffness was considered. In most unfilled polymers, part stiffness can be quite accurately predicted using modern structural analysis techniques and stress–strain data measured with standard tensile coupons. Since polymers are often somewhat nonlinear in their stress–strain relationship, multilinear representations of this relationship may be desirable for extreme accuracy. Furthermore, as we discussed and illustrated by example, there will often be instances when the low values of elastic modulus and high strains to yield or break necessitate the use of analysis tools capable of handling large rotations. Fortunately, the capability to handle both material nonlinearity in the stress–strain behavior as well as kinematic nonlinearity due to large rotations and strains is increasingly available today and is becoming ever more efficient to apply.

However, there are classes of plastics where the relationship between mechanical material property measurements and prediction of part stiffness are not as well established. In many cases, part of the complication has to do with innovative processing techniques that closely couple the part shape to material properties. The discussion of structural foam materials in Chap. 4 is a good example of such a situation. In this case, measurements of the elastic properties using standard molded tensile coupons are not usually accurate because the geometry of the tensile coupon creates a material morphology that is different from that usually encountered in actual parts. Enough work has been performed in this area to understand how useful material properties can be measured and how finite-element analysis using layered shell elements can be used to apply such data to predict performance in real parts. In addition, the relationship between local material density of these foam materials and the material's engineering stiffness has also been explored. Such information has allowed for more accurate design analysis with these materials, but there are still many unresolved issues. In most foam parts, the density of the material will vary, and, as a result, so will the elastic constants. To date, there is no proven technique to predict local density, and hence local engineering stiffness, as a function of the part geometry. As a result, average density values must be used as a first approximation.

Another good example of a plastic material that still needs fundamental attention with respect to its stiffness characterization is the im-
pregnated, random, long-glass-mat thermoplastic (GMT) composites.
In Chap. 4, emphasis was placed upon the fact that these materials
possess a "microstructure" with a scale much larger than that with
which engineers are generally accustomed. As a result, standard extensometer measurements of strain lead to large variation in the measured elastic modulus of GMTs. This is only one complicating aspect of
such materials. As also briefly discussed in Chap. 4, these materials
are usually processed via a compression-molding process that often involves "flow" of the material in the mold. This flow process can lead to
both fiber-volume content variation and preferred orientation of the
glass mat. Neither of these effects is currently well understood now, yet
both are related to the consequent effects upon elastic properties. In
addition, this type of behavior also calls for an ability to predict physical morphology and property changes resulting from the part processing much like that encountered with the thermoplastic foam.

There are other forms of thermoplastic composites that require attention with respect to stiffness issues. Short fibers that are added to
thermoplastics used in the injection-molding process extend the properties of a resin and also increase the complexity of plastic part design.
The nonhomogeneous microstructure that develops during molding
causes mechanical properties to change continually throughout the
part. This variation should be accounted for with fiber orientation and
micromechanics models—which are not yet well established—and used
in the structural design of plastic parts.

Thermoplastic foam, GMTs, and short-fiber injection-moldable thermoplastics are just three examples of new types of plastic materials
that will require additional attention to issues involving the relationship between processing and material stiffness. As material and process inventions continue, more rather than fewer of these situations will
likely arise.

Unfilled thermoplastic materials will probably always have low elastic stiffness values. As a result, innovative structural design coupled
with new processing techniques will be utilized to expand the range of
engineering applicability of these materials. In cases such as the foam-
filled and tacked-off blow-molded panels, engineering modeling can
play an important role in the understanding of geometric parameters
that contribute to part stiffness, allowing designers to optimize performance relative to these values. Once new subcomponents have been
designed, it may be necessary to develop accompanying analysis techniques that can adequately and efficiently analyze the subcomponent's
contribution to the performance of larger structures.

The subject of strength, discussed in Chap. 5, is a more difficult engineering issue to deal with for most materials, and plastics are no ex-

ception. As we saw in this chapter, there are many thermoplastics for which the constitutive theory of plastic deformation seems to be quite appropriate. A great deal of progress has been made in the understanding of the physically observed phenomenon of propagating necks seen in many plastics, such as polycarbonate. However, there are also plastics that exhibit nonlinear behavior that does not appear to be entirely consistent with all the expectations associated with the theory of plasticity. M-PPE, for example, reaches a limiting stress and then incurs additional strain without significant increase in stress. The most important inconsistency between M-PPE's observed behavior and expectations based on plasticity theory is that no localization (necking) occurs. Furthermore, although internal voiding is recognized as being a significant mechanism in the observed nonlinear deformation of M-PPE, there has been no reported work associated with application of any constitutive theories to account for this behavior fully. As a result, plasticity is usually applied to describe the behavior of this material, but the consequences of its inconsistencies on failure prediction are not fully understood.

Another example of a material where proven failure criteria are still lacking is the aforementioned glass-mat thermoplastics (GMTs). This class of materials is usually quite brittle in a tensile test, exhibiting relatively low strains to failure for a plastic. However, in more complex stress states, parts made of this material appear to be able to reach strain levels well beyond the tensile ultimate strains without catastrophic failure. As a result, failure prediction of an engineering component made of GMT can be quite inaccurate. Foamed plastics pose similar problems, although they usually exhibit somewhat larger values of strain-to-failure in tensile tests than GMTs. Established engineering approaches to predicting failure for components made of this class of materials are also still lacking.

Notch sensitivity in thermoplastic materials is another failure phenomenon discussed in Chap. 5 that will require more attention before it can be effectively implemented into engineering design. Brittle failure of thermoplastic components that are quite ductile in tensile tests still occurs too often. Although there is a general understanding of the relative notch sensitivity of different materials, it is not generally possible to determine whether a particular geometry will fail in a brittle fashion at a particular temperature.

Although a number of issues associated with impact response were discussed in Chap. 6, brittle failure of ductile thermoplastics during impact is perhaps the most important outstanding engineering issue that still needs significant development. Although the discussion in Chap. 5 centered on the geometric effects controlling brittle failure due to notch sensitivity, higher rates of loading encountered during impact